TECHNOLOGY INTEGRATION THROUGH AGRO-ECO SYSTEM ANALYSIS

By the same authors

Participatory Rural Appraisal in Agriculture and Animal Husbandry : A Training Manual

---

About the Authors

Dr (Miss) Shagufta Jamal is a Reader in Department of Adult and Continuing Education and Extension, Jamia Millia Islamia Central University, New Delhi. Earlier she worked as Research Associate in a Ford Foundation-funded Project; Teaching Associate at G.B. Pant University of Agriculture and Technology, Pantnagar; and in Division of Extension Education and at Krishi Vigyan Kendra (KVK), Indian Veterinary Research Institute, Izatnagar.

Dr. Jamal has been the recipient of University Merit Certificate Award during 1975-76. She has been honoured with the Young Scientist Award in 1998 and National Award named ISEE Fellow Award in 2003 by Indian Society of Extension Education for her outstanding work in extension research and field extension activities.

Dr. Jamal has to her credit a book, *Women in Dairy Development* and over three dozen papers. She has also presented two dozen of papers in national and international seminars and conferences.

Dr. H.P.S. Arya is the Head of Division and Joint Director (Extension Education) at Indian Veterinary Research Institute, Izatnagar, Bareilly (U.P.) Starting his professional career in 1963 as Professor & Head, Department of Agricultural Extension at one of the Agricultural College in UP, he has worked in various organizations and in various capacities during his career of 40 years.

Dr. Arya has worked on various country-wide projects, such as Communication in Development Administration, Functioning of Village Level Workers, Study on Satellite-T.V. Experiment while working at NIRD, Hyderabad. While working at IVRI, he has worked in the World Bank funded projects including NATP-Project on Collection, Documentation and Validation of Indigenous Technical Knowledge (ITK) as Zonal Leader. He has visited Australia and Japan in connection with training courses and study mission.

Dr. Arya has published six books, more than 60 research papers, more than 100 technical bulletins and leaflets. He has presented more than 50 research papers in national and international seminars and conferences. He has been the recipient of ISEE Fellow Award of Indian Society of Extension Education.

TECHNOLOGY INTEGRATION THROUGH AGRO-ECO SYSTEM ANALYSIS

Using Participatory Approaches

SHAGUFTA JAMAL
H.P.S. ARYA

CONCEPT PUBLISHING COMPANY, NEW DELHI-110059

ISBN 81-8069-119-5 (P.B.)

First Published 2004

Published and Printed by
Ashok Kumar Mittal
Concept Publishing Company
A/15-16, Commercial Block, Mohan Garden
NEW DELHI - 110059 (India)

Phones : 25351460, 25351794
Fax : 091-11-25357103
Email : publishing@conceptpub.com

Preface

India has passed through a number of phases in respect of approaches and strategies of rural development. During fifties, considering the farmers as ignorant, illiterate and orthodox, efforts were made to teach them for change and to adopt scientific technologies. This provided a ground for sixties when the country got certain high yielding varieties transmitted to farmers for adoption. This led to Green Revolution, but only sectoral and only in the green areas with full of resources. This led to widen the gap between haves and have-nots. The so called new technology was found inapplicable to vast under- and undeveloped areas of the country so also for small and marginal farmers who were resource poor. The institutional reform to strengthen its agricultural extension services through 'Training & Visit' system and also several Operational Research Projects have confirmed that the technologies produced at research station are inappropriate for the field situation specially under resource poor condition.

There has been a realization that the agricultural research and extension system should become more farmer-relevant, farmer-oriented, farm-applicable and farmer driven. Recognizing the fact that all the scientific technologies are not relevant and applicable to all the situations, the Indian Council of Agricultural Research launched several programmes for field testing of technologies. Experiences on farming systems research have further led to realize the importance of increased farmers' participation in technology testing, refinement and integration in their farming system. A Pilot Project on Technology Assessment & Refinement through Institution Village Linkage Programme (IVLP) was launched during 1995.

Close involvement of the authors in the IVLP during 1995-98 and in several other programmes of agricultural development having farmers' participation has given them the opportunity to be directly associated with a range of experiences, processes, interactions, methods, perspectives and action and to analyse more critically the application aspects of participatory approaches in technology development, refinement, transfer and integration into the farming systems. The positive feed back received on the work of authors encouraged to write this book which is mainly based on their practical research experiences.

The book explains and illustrates major participatory methods as learning from local farmers about their agro-eco systems, identify problems and their prioritization, problem-cause analyses conducting on-farm-research, assessing and refining the technology, dissemination and integration of technology in the farming system.

The book is divided into nine chapters. Chapter 1 starts with science and technology, its role in development and their interrelationship from the point of view of application of science in the farming activities.

Chapter 2 examines the present agricultural scenario specially after technological breakthrough during sixties. It enumerates the challenges in respect of its development and transformation process. It lists the experiments in respect of approaches analysing the 'transfer-of-technology' (TOT) model and its shift to 'farmer-back-to-farmer', 'farmer first and last' and many other farmer participatory approaches.

Chapter 3 introduces the concept of agro-eco system analysis. Emphasizing its need in agricultural planning and development, the process and methods of agro-eco system analyses have been explained in details. The chapter contains the patterns of analyses, viz. space analysis, time analysis and flow analysis.

Chapter 4 discusses the Rapid Rural Appraisal (RRA) and Participatory Rural Appraisal (PRA) techniques, their concepts, objectives, principles, process and methods. It lists important points for effective application of PRA techniques in rural appraisal.

Chapter 5 discusses various tools of PRA. The tools are discussed first in respect of methodology of their use followed by examples and illustrations. The illustrations are the cases of actual use of PRA tools by the authors and have been presented in the form of figure as well as its interpretation. The tools included are mapping-social map, enterprise map, land use map, cropping pattern, livestock population, transect walk, time analysis—historical timeline, crop timeline, animal husbandry timeline, seasonal analysis—work load, fuel availability, fodder availability, pests and diseases in crops, seasonality of animal diseases, changes and trends, flow analysis—availability of agricultural implements, productivity of different crops, animal productivity, cost benefit analysis of different crops, cost benefit analysis of milk production, matrix ranking—matrix ranking of various crops, matrix ranking of paddy varieties, matrix ranking of fuel, matrix ranking of livestock and quantifying indigenous knowledge, problem cause analysis, identification of intervention point, technology assessment and refinement, technology dissemination and participatory technology development.

Chapter 6 discusses the importance of participatory research and participatory technology development. It discusses various typologies and types of farmer participation in research. Finally, it gives the steps followed in participatory technology development.

Chapter 7 has been devoted specifically on participatory research for women empowerment. The process has been explained by a case of women empowerment through participatory research on improved breed of poultry under backyard system.

Chapter 8 discusses the importance of participatory planning alongwith the methodology followed. It lists important points to be considered and steps followed in the process of participatory planning. It includes identification and prioritization of problems, site characterization, identification of intervention points, technology assessment and refinement, technology modules, On-Farm-Research, On-Farm-Trial and Demonstration.

Finally, Chapter 9 deals with technology dissemination and technology integration. Technology dissemination emphasizes training, farmers' visit to research centres and consultancy by experts.

The book is meant for all those who are directly or indirectly involved in planning, developing and execution of agricultural development programmes. For the researchers, it provides methodological framework for participatory rural appraisal, farmer participatory research and participatory technology development. For extension workers it provides practical hints on the use of PRA tools and selection of appropriate technology for use in the field including the methodology of participatory planning. The book could be useful for students in this field for understanding the concepts and methodologies of Participatory Rural Appraisal, participatory research, technology assessment and refinement and Participatory Technology Development, dissemination and integration into the farming system.

There are a large number of individuals with whom we

had intensive and active interactions on the subject during the last ten years. The list is a long one and we regret our inability to acknowledge all of them by name. The manuscript of this book has been written by us during the last five years which provided rich opportunity and long previlege of interaction on the subject with farmers, livestock owners, extension workers, subject matter scientists, social scientists, extension scientists, planners and policy makers, teachers and students. In this context we profoundly thank to Dr. P. Das, Dy. Director General (Agril. Extension), and Dr. S.P. Singh, Ex-Principal Scientist, Indian Council of Agricultural Research, New Delhi. We are thankful to the farmers of the Project villages who were the real inspirers for compiling our experiences.

We must thank Shri Ashok Mittal of Concept Publishing Company not only for his keen interest in publication of this book but for his constant encouragement and personal motivation in completing this work.

Last but not the least we express our gratitude to our parents and all family members who have been our constant source of inspiration.

Shagufta Jamal
H.P.S. Arya

Contents

List of Figures

List of Tables

1
Science and Technology

Science is the systematic study to discover the fact and phenomenon of the nature in the form of bits of knowledge. These bits of knowledge when put together in application form is considered as technology. Our farmers have long traditions of farming who have been following certain practices without any scientific reasoning, testing or study. Technologies are being developed and integrated in the rural farming system through experiencing the applications and their consequences without having the basic knowledge of facts and phenomenon of the nature through systematic study. On the other hand our agricultural scientists are stirring hard to develop agricultural technologies based on scientific information derived through systematic studies.

A large amount of knowledge has been generated through research and a great number of technologies have been developed for their application to increase production and productivity by the agricultural scientists. Efforts have been made to take these technologies from the research stations to the farmer's fields for their application. In spite of a huge investments on transfer of technology programmes in terms of money, national man power and infrastructure development not more than 30% of the generated technology has been reached to the farmers and used for increasing production.

Technology

After independence, our planners, development workers and

the scientists felt that the farmers are ignorant of scientific technology and that if the scientific technology generated at the research station is transferred to the farmers in the convincing way, its application can increase the production. Later during seventies and eighties it was realised that the technology developed at the research stations can not be applied in the existing farming situations under rural setting due to the limitations of infrastructure and other micro level factors. This made to think for the efforts to develop infrastructure resources at the farmers level. Soon it has been realised that in no way it is possible to develop the farmer's resources at par with research stations where all the factors of production can be brought under human control. This is specially true in case of small and marginal farmers who are the bulk of the farming community.

It has now finally been realised that the technology generated at the research station is not suitable and appropriate for its application to the situation of our farming systems. This is the reason why 70 percent of the generated technology has still not been applied in the field. This challenges the technology development process and the existence of research system.

The technology development process needs to be analysed critically and modified in such a way so that it could yield technologies which could be applicable to the existing farming situations. Technology development must be based on the continuing generation of knowledge, selection of ideas, their proven utility and on their assimilation and integration in the farming systems with the farmers which is transferred through generations and cultures.

The agricultural development involves the adaptation and control of environment. The extent and nature of development moves in the following manner between the adaptation-control continuum. The utilization of knowledge in the form of technology is an important instrument for increasing control. When development takes place the line AB moves from left to right and control increases.

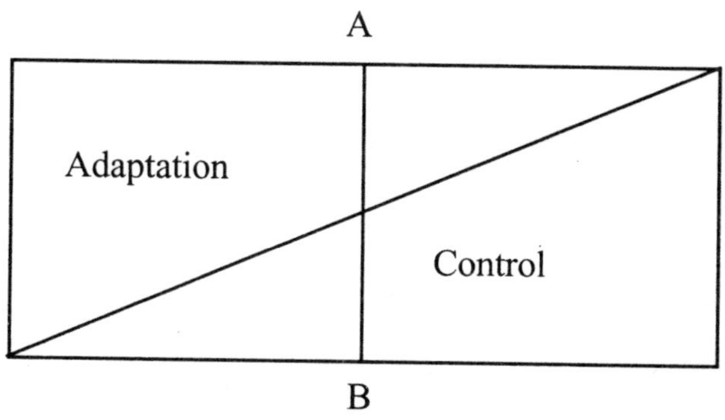

Technology development is groomed by this continuum. It has to have a realistic balance between the control of and adaptation with the environment, Technology development, therefore, will have to vary from situation to situation depending upon the extent of sustainable control over the environment. Illustratively, technology development at research stations and at organized farms will not hold good with the farmers situations specially with the small and resource poor farmers where the control of environment is negligible and the technology needs to adapt with the environment.

The discovery or development of new knowledge by the scientists are usually in the form of pure knowledge (theories and principles) which is transmitted by verbal or written words. This is normally and essentially the function at the research stations and in the laboratories. Whereas, the technology development process directs its attention to the users of technology. Technology development refers to bring the bits and pieces of generated knowledge together for the users (farmers) in their own situations, controlled, semi controlled or uncontrolled.

2
Agriculture Scenario and Challenges

Agriculture is the back bone of Indian economy. For ever rising population, the demand for agricultural production is increasing. After independence, during the fifties, the peasants were liberated from the stronghold of exploiters, land was titled to the tiller and foundation for modern agriculture was laid by way of investments in irrigation and fertilizer. The sixties and seventies witnessed the development and widespread adoption of the seed-irrigation-fertilizer - plant protection technology ushering in the green revolution. Programmes were initiated also in drought prone and rain fed areas. Simultaneously programmes of rural employment generation and poverty amelioration were launched. Research efforts were also intensified towards development of dairy, poultry, piggery, fishery and horticulture including quality improvement programmes in different sectors.

While the process of consolidation of the green revolution continued in the eighties and nineties, research was strengthened in the fields of pulses, oilseed, integrated pest management, conservation of natural resources and their effective utilization. The process of rural development through various integrated rural development programmes continued to reach and spread among small and marginal farmers, agricultural labourers, rural artisans, weaker sections of the society and even the rural women. The nineties have brought into focus the twin objectives of enhanced productivity and sustainability.

Notwithstanding the impressive gains in agricultural production, the vast agricultural potential still remains highly under realised. There are serious gaps both in yield potential and technology transfer as the national average yields of most of the commodities are very low. Till such time the population stabilizes, the demand for food and other needs would be on the increase depending upon income-oriented consumption pattern. Fast declining input use efficiency in major cropping systems, rapidly shrinking resource base and non-availability of favourable factors of past growth make it difficult to keep pace with the present rate of population growth and consumption pattern.

During the 21st century, agricultural development in the country would be guided not only by the compulsion of improving food and nutritional security, but also by the concerns for environmental protection, eco-friendliness, sustainability, stability and profitability. Globalization of markets would call for competitiveness and efficiency of agricultural production. In the coming years, agriculture will face challenging situations on the ecological, global, climatic, economic, equity, social justice, energy and employment fronts.

Fortunately, there have been several technological breakthroughs in the recent years and new cutting edge technologies are appearing regularly. These provide new opportunities and augur will with the challenges posed. Developments in biotechnology and molicular biology, ecology, system analysis and crop modeling, socio-economic analysis and management science and informatics provide new potential.

There is an increasing awareness of the role of agricultural development not only as a supplier of food but also as a driving force for overall economic growth, employment generation, poverty alleviation, food security and sustainability. Therefore, an agricultural based development strategy relying on increase in the productivity, specially of small land holders and the provision of employment opportunities in the farm and non-farm sector is intrinsically poverty reducing and food security promoting.

For an effective management of change, problem-oriented demand-driven, viable, sustainable and diversified system and situation-based matrix approach of integrated research and development involving all reasonable players as partners are believed to offer practical solutions. Research, technological change and dissemination are critical elements of agriculture-based development strategy. There are serious gaps in the agricultural research and technology development system.

Technology Transfer

The public sector research and technology transfer systems in India have been highly effective in introducing the green revolution technologies to farmers, but only in well-endowed areas. The overall efficiency of the system has been rather low. The technologies have been less successful in poorly endowed areas and several second generation problems, specially unsustainability, have emerged. Technical solutions to these new challenges are expected to be more locally based. Their solution will depend on a system approach and requires a bottom-up response to the demands of technology users. Closer interaction between research, extension and farmers through participatory approaches will be needed.

The Transfer of Tecnology (TOT) model is deeply embedded in the thinking of many professions and disciplines around the world. It is the structure with centralised knowledge, in which power, prestige and professional skills are concentrated in well informed 'cores' and 'centres'. These cores and centres generate new technology which then spread to the peripheries. Highly trained medical scientists, civil-mechanical and agricultural engineers, agronomists and others develop technologies in laboratories, workshops, experiment stations and then attempt to transfer them to would be clients. This approach has had immense successes in industry and agriculture with resource rich clients. For example, the development of mechanization through combine harvesters, tractors and threshers by agricultural engineers and the development of high yielding technology packages by plant breeders and others have enabled many of the resource -rich to increase their productivity and profitability. But, this approach has had several shortcomings for other would-be clients who are resource-poor.

In most agricultural sciences, the centres in which research is conducted are experimental stations, glass houses and laboratories supported by super back-up services with provision for controlled conditions, with excellent access to inputs without significant cost of labour constraints and without the requirement that a crop must be marketed and make profit. Scientists in experiment station, glasshouses and laboratories generate or test the new technologies and then pass them over to extension services to transmit to farmers.

In practice when the resource poor farmers did not adopt so called "good" new technology, both agricultural scientists and social scientists at first attributed this to ignorance of farmers. The large scale social scientists in India in the 1960s on 'diffusion of innovations' assumed that the technologies were good and appropriate. A major premise was that, if small farmers did not adopt them, it was because they did not know about them or did not know enough about them. The prescription that followed was for more and better extension "to educate the farmer." This exactly reflects the pattern of thought, i. e. "We have the relevent knowledge which ignorant farmers do not have. We must teach the ignorant farmers."

But, there is now much evidence and understanding that when resource poor farmers do not adopt technology it is usually not from ignorance but because the technology does not fit their needs and their physical, social and economic condition. Technologies whether biological or physical, bear the imprint of the conditions in which they are generated. They are adaptable easily in similar conditions but often not adoptable where conditions differ. As it happens, many conditions on research experiment stations and in laboratories are sharply different from those of resource-poor farmers.

There are strong professional reasons why agricultural scientists follow the so called "TOT" Model which does not fit to the resource-poor farmers and for India where majority of farmers are resource-poor. At the international and national levels there is the prestige attributed to 'high' technology, seed breeding and expensive and sophisticated equipment and methods of research. There is personal convenience in working in office and laboratory and on a research experiment stations, rather then on farm or with farmer, Further, for gaining professional recognition and for minimising risk of not gaining it through failed experiment work in laboratory and on research station having controlled environments is preferred. The environments of resource poor farmers are very complex. There are many stresses with too many interaction (ifs and buts). Moreover the research methodology for such environments is not well established and standardised. It is safer for professional advancement and recognition not to share the farmer's risk.

Scientists motivations interlock with well known biases of professional behaviour, contact and perception towards those rural people who are better off and to neglect those who are poorer. Scientists are often urban based. Their rural visits have spatial biases-urban, tarmac and roadside and towards large villages concentrating where better - off tend to be located. Other biases concern contact with those with higher status, more influence, greater wealth, better education and better living facilities and standards. Scientists meet adopters more than non-adopters. Demonstrations are laid out most often on the land of such resource - rich farmers who provide hospitability and cups of tea for visiting officials. Scientists have biases of modernity and capital - intensity : it is the tractor, the pump, the thresher, the inorganic fertilizers and other purchased inputs, which attract attention. They are also season - proofed in that they do not personally experience the vagaries and difficulties of weather, i.e., heavy rains, flood or drought as farmer does. Their pay cheques are regular, monthly and equal, not seasonal and variable as in case of farmers depending on uncertain agriculture.

In the light of disappointing experience with transfer of technology to resource-poor farmers, many modifications have been made. Some of the changes have taken the form of organizing feed back to researchers on problems in adapting and adopting the recommendation. The T and V provides for feed back from extension to research through strong linkage (Benor, D. and Barter, M. 1974). IRRI's constraints research is another example (De Dutta S.K. et. al. 1978) where yield gaps were measured between performance on the research stations and on the farmers' fields. The Operational Research Project in India also illustrates this pattern which provides scientists with opportunities to test, verify and perfect the new generated technology at the research station while it is operated under field conditions. It is not experimentation but only a step to verify the results of successful experimentation conducted elsewhere. In all these three examples - T & V, IRRIs' constraints research and the ORP in India - the research comes first to develop the technology which may then be adapted and perfected following experience with its use in on-farm conditions. Despite modifications only for feedback, the basic TOT structure remains unchanged (ICAR 1995).

The T O T model and modifications are well exemplified in major agricultural research programmes in India. Research on specific area problem is conducted through All India Coordinated Research Projects located in Agricultural Universities and ICAR Institutes. Operational Research Projects were implemented for specific problem areas. For small and marginal farmers and landless agricultural labourers, the Lab-to-Land' programme was initiated. The major thrust was the introduction of new technologies generated at the research stations. These programmes present progressive modifications of the model and attempt to offset its biases. There has been increasing emphasis on on-farm trials and demonstrations.

The National Demonstration programme was implemented to test and demonstrate the yield gaps between research stations and the farmers' fields along with the potentiality to narrow the gap. The establishment of Krishi Vigyan Kendra in backward areas for training farmers and the 'Lab-to-Land' programme were directed to problem environments focuses on farmers who are often resource poor.

The outcome of these modifications remained disappointing. The old explanation of 'ignorance' on the part of the Resource Poor Farmers (RPFs) was partly superseded by attempts to understand farmers' conditions and constraints. But, the basic model remains the same. Priorities are set by scientists relying on their professional interest, understanding and criteria. Research is conducted on central location and then tested, advocated and extended outwards. There has, it is true, been increasing emphasis on feed back from the field. But, the farmers from whom there is feed back tend to be precisely those best placed with resources. It is scarcely to be expected that RPFs who always leg behind in taking benefit of technology will provide feed back. They do not come forward and speak up to scientist and the scientists are also least able or inclined to learn from them. To conclude, the TOT model does not encourage scientists to learn from RPFs.

Alternatives of TOT models - Experiences else where

The Farmer - First and Last (FFL) model advocated by Chamber, R and Ghildyal, B.P. (1985) entails fundamental reversals of learning and location. FFL differs from TOT in starting not with scientists and their perceptions and priorities, but with RPF families and theirs. It begins with systematic process of scientists learning from, and understanding RPF families their resources, needs and problems. The main focus of research and learning is the resource poor farm rather than research station and laboratory, The research station and laboratory have a referral and consultancy role, secondary to, and serving the RPF family. The criterion of excellence is not the place of research but the test of whether the new practice spread among the resource poor.

Farming Systems Research is a prototype of FFL which begin with systematic attempts to understand the farm and farming system. Many have given the meaning to 'Farming

Systems Research' as a research wherein the elements of a farming system are taken into account and investigated on an experiment station. This is not FFL but a TOT approach.

Farming Systems Research (FSR), in its various manifestations, is often described in terms of stepwise sequences. Shaner et al. (1982) emphasize five activities - (i) target and research area selection, (ii) problem identification and development of a research base, (iii) planning on-farm research, (iv) on-farm research and analysis, and (v) extension of results, with collaboration between these and extension and the experiment station. Maxwell with reference to Norman lists activities in a slightly different classification as to identify recommendation domains, diagnosis, the generation of recommendations, implementation and monitoring and evaluation.

The sequence of activities, however, could vary by circumstances. The quickest and most cost-effective approach might be - inventive, opportunistic and iterative, not necessarily following a fixed order of activities. According to Rhoades (1982), in the farmer-back-to-farmer approach, there is more flexibility in methodology. We might even start by conducting experiments with farmers just to learn about a problem,. This approach believe in the rapid appraisal methodology (informal), but it even uses the sondeo in evaluating impact. Rigid, step-wise field methodologies never worked.

Some prototypes of FFL approach have been experimented such as CIMMYT's approach to planning technologies appropriate to farmers, the Sondeo method of rapid appraisal, ICRAF's diagnosis and design (D & D) for agro-forestry and farmer - back - to - farmer methodology of CIP.

The CIMMYT approach emphasises the farmer as the primary client of agricultural research and farmers' circumstances as the basis for planning research. Farmers are grouped into 'recommendation domains,' groups of farmers for whom more or less the same recommendations can be made. Rapid appraisals are conducted by an agronomist and an economist working together. This is followed by exploratory survey and then by formal verification survey. On the basis of the available data prescreening of technologies are done to identify 'best bets' for on-farm experiments.

The Sondeo approach developed by Hildebrand in Guatemala is strongest in its technique for the creative combination of disciplines in rapid appraisal to generate new technology. A zone with homogenous farming practices is identified, in which farm trials of technologies are conducted. A team leader and ten team members, five of them agronomists and animal scientists and five from socio-economic disciplines conduct a very rapid appraisal. They work in pairs - one agronomist or animal scientist with one socio-economist, changing partners each day for five days. They visit the area and interview the farmers and others attempting to understand the farming system and to identify feasible and suitable improvements, and all brainstorm together each evening. At the end of five days, many three cornered discussions between farmers, social scientists and biological scientists have continued to proposals for improved farm practices. A report is written under pressure and provides proposed innovations for the Technology Testing Team which then works in the area with on-farm and with-farm trials.

ICRAF's diagnosis and design (D & D) methodology sets out to identify promising agro-forestry technologies. Major emphasis is placed on the farm household management unit and the satisfaction of its needs. A minimal team includes one or more representatives of agricultural science (agronomy, horticulture and linestock sciences), forestry, social science (sociology, anthropology, human geography and economics) and natural sciences concerned with land resource survey (ecology, soil science and climatology). The application of D and D procedures by a multidisciplinary team usually entails about two weeks to carry out the diagnostic survey, analyse the results and develop appropriate design concepts for agro-forestry interventions to improve the existing land use system. There is a four stage procedure - prediagnostic, diagnostic, design and follow up planning.

The original 'farmer-back-to-farmer' research was conducted on potato storage in Peru by biological scientists and an anthropologist following 25 years of failure in potato storage work. The anthropologist learnt about farm families' objectives and their knowledge of, and problems with potato storage and acted as a link between them and the biological scientists, bringing the later into direct learning contact with the farmers. There were four stages - establishing a common definition of the problem, interdisciplinary team research seeking a solution, testing and adaptation of the proposed technology on-farm, with farmers contributing ideas and 'farmer evaluation - the last judgement'. The result was an improved and adoptable technology which meets farmer's objectives, used materials to which they had access, fitted in with their traditional house design and, above all, was adopted by them. A key element was changes of perception and priority on the part of the scientists. In the beginning, what appeared losses to the scientists were not necessarily losses to farmers, who, in fact, had uses for the shriveled or spoiled potatoes.

Reversals of Explanation, Learning and Location

The major reversal of TOT approach is that *explanation* of

non-adoption shifts from deficiencies of the farmer and the farm level, to deficiencies in the technology and in the technology generating process. The reversal of *learning* requires that scientists start by systematically learning from farmers, with transfer of technology from farmer to scientist as a basic and continuous process. The reversal in *location* requires that R and D take place on-farm and with-farmer, with research stations and laboratories in a referral and consultancy role.

The reversal changes thrust from resource rich farmers to resource poor farmers. The scientists who have been working in the laboratories and on-stations for defining and studying the problem have to shift to on-farm and with-farmer. The shift has been depicted in the following diagramme.

Learning from the Farmers

The changed understanding that the resource - poor farmers are not adopting the new technology due to their ignorance to that the technology does not fit the needs and the physical, ecological, social and economic conditions denotes the ignorance of the scientists about these. Under the circumstances and with such perception, the process has to start from scientists' learning of the farmers and the farming systems. The methodology for systematic, meaningful and timely study has been attempted and standardized through agro-eco-system analysis using various tools of research.

Shift of Activities

TOT		Reversal	
Resource-Rich	Resource-poor	Resource-Rich	Resource-poor
Scientists define problems & opportunities ↓		Transfer of scientists →	Scientist learn from farmers about needs, resources and priorities of farm families. ↓
On-station research ↓			Joint definition of problems and opportunities. ↓
New High-Yielding Technology ↓		On-station referral of problems ↔	On farm, with farmer R & D ↓
Demonstrations and testing on farm ↓			Farmers test and evaluate ↓
Other Resource rich farmers	→ Resource poor farmers	Resource rich farmers ←	Other resource poor farmers

3
Agro-Eco-System Analysis

Agro-eco-system analysis provides an ecology based frame work for understanding farming systems and identifying problems by land type.

Agro eco system analysis (Conway 1985) is concerned with the agro eco system of an area whose micro unit is generally the village. Agro eco-system represents a system comprising of a set-of natural conditions (climate, topography, soil etc.) and a set-of agricultural activity existing to utilize that system. In an agro eco-system analysis a secondary data review precides the field visit where a set-of diagrams are produced on the agro eco system based on direct observation, semi-structured interviews and other visual methods of Rapid Rural Appraisal (RRA). The diagrams produced through field visits are then analysed, on the basis of which a list is made of important factors influencing the agro eco-system in terms of productivity, stability, sustainability and equity. It also gives rise to a set of key questions and related hypotheses which the group agrees to be important. This is followed by an assessment of impact of innovations and interventions on system properties and their prioritisation through ranking. The key questions are both research and development questions where the first category leads to further research while the second category leads to development action (McCracken, Pretty and Conway, 1988).

The methodology of rapid agro ecosystem zoning (Conway et. al. 1987) consists of treating areas in terms of

General Recommendation Domains (GRDs) to help in development planning. GRDs consists of individual domains or areas characterised by similar biophysical and socio-economic features on the basis of which appropriate recommendations are made regarding innovations and interventions.

The field oriented methods have some features in common such as (i) quicker than conventional methods. (ii) applied in field situations, (iii) learning directly from people (iv) semi structured, multi disciplinary approach with flexibility and innovative and (v) emphasis is on understanding processes through involving people, gathering insights on problems and their probable solution rather than final answers and recommendations (Conway, 1988). However, RRA and similar methods remain extractive i.e. researchers observing, writing and analysing. The field oriented methods leave much scope for inducing greater elements of participation.

Agro eco system analysis is a new technique of analysing the problems and finding solutions to the problems related to agricultural development. There is now a growing realisation that the problems of agricultural technology can not be tackled individually in an isolated manner. These problems are systemic in nature and are linked with each other due to varying agri-ecological and socio-economic processes.

A system is defined as a sum total of elements contained within a boundary having a strong functional relationship with each other. The combined out come of the functional relationship within the boundary is to produce a distinctive behaviour such that it depends to stimuli as a whole even if the stimulus is only applied to one part.

Properties of Agro-Eco-System

The essential behaviour of agro-eco-system is characterized by four system properties. These are : productivity, stability, sustainability and equaitability. These properties are relatively easy to define but not equally easy to measure. These properties can be defined as :

1. Productivity : It is the net increment in valued product per unit of resources (land, labour, energy and capital). It is commonly measured as the annual yield or net income per hectare or man hour or energy investment.

2. *Stability* : It is the degree of productivity which remains constant inspite of normal real fluctuations in environment variables such as climate or in the economic conditions of the market. It is most conveniently measured by the reciprocal of the coefficient variation in productivity .

3. *Sustainability* : It can be defined as the ability of a system to maintain its productivity when subjected to stress or perturbation. A stress is defined as a regular, sometimes continuous, relatively small and predictable disturbance, e.g., the effect of growing soil salinity or indebtedness. A perturbation is an irregular, infrequent, relatively large and unpredictable disturbance such as drought or flood or a new pest. The measurement of sustainability is difficult but can often be done retrospectively. Lack of sustainability may be indicated by declining the productivity or sudden fall in production.

4. *Equitability* : It is a measure of how evenly the productivity of agro-eco-system is distributed among its human beneficiaries. The more equitable system, the more equally distribution of agricultural produce or income or resources shared among the population of farms, village, region, or a nation.

Procedure of Agro-Eco-System Analysis

The analysis must be based on the following four assumptions:

(i) It is not necessary to know everything about an agro-eco-system in order to produce a realistic and useful analysis.

(ii) Understanding the behaviour and important property of a system requires knowledge of only a few key functional relationship.

(iii) Producing significant improvements in the performance of a system requires changes in only a few key management decisions.

(iv) Identification and understanding of these key relationship and decisions require that a related number of appropriate key/problems questions are defined and answered.

Keeping in view the above assumptions, three patterns are chosen to reveal the key functional relationships that determine the properties of an agro-eco-system. These are space-analysis, time analysis and flow-analysis.

Space Analysis : The spatial patterns are most readily revealed by simple maps and transects. Transects are useful in defining a system boundaries and in identifying problem area.

Time Analysis : The patterns in time are best expressed by simple graphs to study seasonal changes, crop calendars, cropping sequences, labour, credit needs, prices etc. This helps to identify the periods in the year when the timing of operations and availability of resources are critical for productivity and stability. The long term changes in prices, production, climate and demographic parameters can also be studied through time analysis.

Flow Analysis : With the help of this analysis one can observe the patterns of flows and transformation of energy, materials, money, information etc. In a given agro-eco-system flows should be analysed for finding out the major causes and effects and also for the reasons of stabilizing or destabilizing factors. The flow analysis can be made with the help of flow diagrams, tables, matrices, bar histograms and regression graphs etc.

4

Rapid Rural Appraisal and Participatory Rural Appraisal

It is of utmost importance to understand the intricacies of socio-economic and ecological environment within which the farmers operate and resolve their problems by collaborating with them according to their needs and priorities. Farming Systems Research (FSR) was evolved in 1965 whose objective was to further research in the area of farming systems, analyse problems and plan research and extension activities. It was evolved as an approach to agro eco system analysis.

Short surveys of conventional analytical approaches became apparent with the spread of Green Revolution technologies as attention spread to the problems faced by farmers with limited resources operating in marginal heterogeneous environments. It became apparent that methods were needed to understand the overall rural systems in which farmers operate and to assess technologies that may fit these systems. This required techniques to produce quicker and more accurate analysis of the complexities of small farm systems and farming systems of the resource-poor. Thus, the methods of Rapid Rural Appraisal evolved.

Rapid Rural Appraisal or RRA developed as a methodology in 1970s, was influenced by FSR and other methods of agro-eco-system analysis. Some of the early path breakers of such methodology were Robert Chambers, Peter Hilderbrand, Robert Rhoades and Michael Collinson who along with others met in conferences at the Institute of

Development Studies during October, 1978 and December, 1979. Through the literature the concept of RRA spread in different parts of the world and in the mid-eightees there was a wide array of experience from the application of RRA in different field situations.

Rapid Rural Appraisal is a way of organizing people for collecting and analysing information within a short time span. It can be defined as any systematic process of investigation to acquire new information in order to draw and validate inferences, hypotheses, observations and conclusions in a limited period of time. It has flexibility to adjust to situations because it does not imply or recommend a standard set of methods to be applied in each case.

Kinds of RRAs

The RRAs can be classified into four kinds on the basis of objective of the investigation -

1. *Exploratory RRA* : These are sets of preliminary key questions and hypotheses. Most of the RRAs are exploratory in nature and focus on farm production system.

2. *Topical RRAs* : Based on the initial findings of the exploratory RRAs, additional questions and hypotheses can be explored in detail focusing on specific aspects. These are called 'Focused RRA' or 'Topical RRA'. Topical RRAs provide information on specific topics such as gender roles, feeding of livestock, milk production, processing and disposal, grain and straw storage etc. Topical RRAs are used to develop hypotheses to serve the basis for research and development.

3. *Participatory RRAs* : This phase involves villagers and local officials in decisions about further actions based on the results of the topical RRAs. This action can result in on-farm trials.

4. *Monitoring RRAs* : These RRAs are used for monitoring the progress of action - trials, experiments, development activities, etc. with the output typically consisting of changes or modifications to improve further performance.

There are some limitations on the use of RRAs. The practitioners many times tend to take a package of preconceived ideas and potential technologies to the field and thus limit the discussion and restrict the scope of the RRAs around that package. Further, the requirement of shifting paradigm of development was not only a relatively quicker method of data collection but also one which would help local people to perform their own analysis. Hence, certain elements were required to be incorporated in RRAs and other field-oriented methods to make them participatory and more broad-based in terms of their applications. It was a shift of emphasis from "rapid" or "quick" to "participatory" and "empowering' which fitted well with a participatory approach to rural development and become the starting point of PRA.

Participatory Rural Appraisal

Participatory Rural Appraisal (PRA) is methodology for interacting with villagers, understanding them and learning from them. PRA is a means of collecting different kinds of data, identifying and mobilising intended groups and evoking their participation and also opening ways in which intended groups can participate in decision making, project design, execution and monitoring.

Because of its participatory nature it is useful methodology to focus attention on people, their livelihoods and their inter relationships with socio-economic and ecological factors.

PRA constitutes a process of involvement-with rural people for indigenous knowledge building exercises. It is a way of learning from and with villagers to investigate, analyse and evaluate constraints and opportunities.

The participation of rural people can be facilitated through PRA for planning, implementing and monitoring of rural development programmes.

PRA started first in Kenya in 1988 and side by side in India. In India, it was started in Agha Khan Research Foundation in Gujrat by Jennifer Akleen. The Action Aid programme in Bangalore did some work on PRA and also Robert Chambers workcd on PRA in Hyderabad.

PRA is sometimes known as Participatory Rapid Appraisal where the emphasis is both on Participatory and Rapid. The emphasis on rapid, however, is more in terms of data collection and less in respect of process of development or even implementation of plans. Other names are also used for describing PRA, some of which are: Participatory Rapid Rural Appraisal (PRRA) Participatory Learning Method (PALM) and Participatory Learning and Action (PLA).

Objectives of PRA:

(i) To have greater and better involvement of villagers for learning about their perceptions, experiences and capabilities.

(ii) To generate information and collection of data for immediate or further use.

(iii) For learning about the impacts of earlier or on-going policies and programs and to frame new ones.

(iv) For estimating trends and ascertaining conditions of the issues at hand.

(v) For validation or cross - checking of data collected from other sources.

(vi) For training of different categories of persons involved in the development process, whether from the government, NGOs, banks, donor agencies, researchers, extension agents, scientists etc. and

(vii) For research studies on the use of PRA and to suggest improvements in its methodology.

PRA is a family of approaches and methods to enable rural people to share, enhance and analyse their knowledge of life and conditions, to plan and to act (Robert Chamber, 1993). It is a way of learning from and with community members to investigate, analyse and evaluate constraints and opportunities.

Development planners and project implementers do the planning for the villagers. Surveyors come with their questionnaire to the villages and extract the data. In such situation the local people have the role of a passive respondent. Consequently, programs come from the top to people making them recipients of help and development fund. In those cases, the people's real need are not taken into account, poor groups are side lined and self-help initiative of local people and self responsibility for the programs is completely over looked and missing.

PRA believes that the participation, empowerment and a sense of responsibility of the people have to start from the very beginning of the planning process. Only those topics are discussed which people feel are relevant for the explanation of their situation. PRA is bottom up planning in which the real problems after a thorough analysis with all village groups are reflected and problem solutions are supported by all.

PRA is a process of participation with the villagers in which rapport building paves the way for them to perform their own analysis and to express themselves whether by means of 'verbals' like narration or visuals such as making a map. The final product of PRA would differ in output and content depending on several factors. Hence, the out put of PRA can not be predicted with any degree of certainty.

PRA Concepts :

* PRA is a methodology for interacting, understanding and learning from the villagers.
* PRA is a means of identifying and collecting different kinds of data in a village.
* PRA identifies and mobilise the intended groups and evoke the participation of people and also opening ways in which extended groups can participate in decision-making, project design, execution and monitoring.
* PRA believes that the participation, empowerment and sense of responsibility of people have to start from the very beginning of planning.

A comparative analysis of RRA and PRA over the years is being presented below:

	RRA	PRA
Period	Late 1970s-1985	late 1980s-1990s
Major Institutions	Universities	NGOs
Main Users	Aid agencies & Universities	Govt. & field organizations
Key Resources	Local people's knowledge	Local people's capabilities
Main Innovations	Methods	Behavior of outsiders
Prominent Mode	Elicitive, extractive	Facilitative and participative
Ideal objectives	Learning by outsiders	Empowerment of local people
Long-term outcome	Plans, projects, publications	Sustainable local action

Principles of PRA :

1. Learning and listening
2. Offsetting biases
3. Triangulation (cross-checking)
4. Sharing
5. Congenial attitudes and friendly behaviour with villagers.
6. Seeking diversity
7. Optimal ignorance
8. Facilitating (villagers do themselves)
9. Self critical awareness and responsibility
10. Reversal of learning.
11. Learning rapidly and progressively

1. Learning and Listening

PRA is based on the principle of learning and listening through participation. Proper learning can take place with appropriate mental and physical set up of a learner. The scientists shall have to be mentally prepared to listen, learn and show respect towards those from whom such learning can take place. Learning with a closed mind becomes a difficult task and fails to take place. With appropriate attitude and behaviour, learning experience can be rapid and progressive.

2. Offsetting Biases

PRA aims at offsetting biases which generally accompany a "rushed" appraisal by professionals for quick results. These professionals or scientists tend to appraise themselves through samples at convenient locations, such as roadside locations, at convenient times when the weather is favourable, consult would-be-beneficiaries (mostly men) who are generally better off and professionally tend to look at those aspects which they think are important. In order to offset such biases, PRA encourages released listening and learning, seeking participation from people who are relatively worse off, visiting remote and interior locations and visiting communities at their convenience (Chambers, 1993). PRA is based on the principle of utilizing precious community time in the best possible way. The farmers are busy in pursuing local livelihood activities and it is important to keep track of their time and learn as much as time permits. The farmers may be requested to spare some of their time and learning should be focused so as to make proper utilization of such available time.

3. Triangulation (Cross-Checking)

The process of cross-checking is an important principle of PRA for minimising errors in the results obtained. Since, there are different sources from which information can be obtained, it is important to cross-check the data for its validity. There exists different ways to check the validity of the data in PRA such as changing methods and tools, locations, timings, groups, teams, etc. Cross-checking or triangulation could be done through the following ways:

- Team composition for approaching communities can be multidisciplinary so that information can be elicited from different perspectives. Teams can have persons from different disciplines, experience, gender-groups, age-groups, etc.
- Validation can be pursued by approaching different groups and individuals either separately or together. This is called triangulation.
- Different participatory methods, in combination, can be used to test and varify data. Semi-structured

interview can be followed by a social map leading into a seasonal calender in order to check the validity of a data set.

- Participant checking is another way to test data with the same set of people who generate original information.
- Parallel investigations by changing locations and time of a participatory session can help in validation of data. Multiple investigation groups can also help in validation.
- Peer group checking is another way to validate data by having periodic meeting with peers directly / not directly involved in the inquiry process (Pretty. 1993).
- Comparing data from participatory method with data from other primary and secondary sources can also help in varifying issues and indicate discrepancies and gaps.

4. Sharing

"Sharing" is dissemination of learning through participation. A joint pool of knowledge is created by learning from communities which needs to be shared with others for further learning. Such sharing from PRA sessions enriches the learning process and highlights local perspectives. It is sharing of knowledge, information, views, technology, materials, time and other resources between communities and outside agencies; amongst community members themselves and among outsiders and others associated with such participatory learning.

Sharing is integral to PRA , without sharing participation is not possible. Unless the local people are willing to share their views and time, the outsider professionals can not learn from them. The more they share, the more the opportunities to learn. Hence, sharing is essential for learning to take place and also for its cross-checking and validation. Sharing can also take place after learning has occurred. Such sharing raises new points, gives new insights, provides opportunities to those left out for stating their views, generate additional information and helps in analysis and validation of data.

5. Attitudes and Behaviour

Attitudes and behaviour of outside professionals/scientists constitute the base of approaching and establishing links with local communities. If attitudes are not geared to those of a learner, the local communities may not respond or easily open up for interaction. Qualities like patience, appropriate body language, appreciation of local expertise and respect for community feelings help in "breaking ice" and building rapport with local communities. Appropriate attitudes and behaviour of professionals tend to create positive response from local people in terms of soliciting their participation.

For participatory sessions, people need to be relaxed and also require to have a reasonable level of trust and faith in intervening agencies from outside. Congenial atmosphere is required for participatory sessions with congenial attitudes and behaviour of the outside professionals.

6. Seeking Diversity

PRA is concerned more with analysis of difference rather than looking for representativeness of results or data collected. It is looking for diverse rural events, different processes and forces explaining various relationship in rural communities from different perspectives. Multiple causes generally lead to a chain of impacts, and for any analysis, greater the diversity, better is the representation of local reality. Information on diverse aspects/situations are collected from diverse situations to seek all possible variation. Information from different focus groups are collected to have complete information of the community.

7. Optimal Ignorance

In order to minimise cost and time, the principle of optimal ignorance is applied by facilitators which means knowing what is worth knowing and knowing enough to sense the purpose and not knowing the rest or not trying to find out more. Proper defining and setting of objectives, planning for interview and controlling the discussion as per plan in this respect is obvious.

8. Facilitating

PRA sessions are required to be facilitating people to gather, think and analyse data and make them able to act upon. It is not extractive but empowering people to act upon themselves. The outside professionals work as referrals and certainly not as teachers or instructors. They are information giver to those who seek for it. They act as facilitator and not as planners or actors for implementing plan.

9. Self Critical Awareness and Commitment

PRA sessions provide opportunity for the community to create self critical awareness, to know themselves more critically, systematically and scientifically, their own problems, constraints, limitations, strengths, weakness, opportunities, threats, etc. for themselves. On the basis of the self-appraisal, PRA further provides opportunity to prioritize their problems and to select the possible alternative action with the help of outside facilitator. For action, the PRA sessions provide a sense of commitment and responsibility among individuals and groups.

10. Reversal of Learning

Under conventional method, the farmers are subjected to learn from the scientists and extension workers, scientists learn from the farmers about the practices they follow with their rationality vis-a-vis their situations. The scientists used to develop technologies generally on the basis of their 5 to 10 years experimentations and experiences in the laboratories and at the experiment farms, while the farmers have developed their practices through experiences gained over generations which consist of number of experimentations. There must be some logic behind those age old practices. The only one thing that makes them so called unscientific, is that the results of numerous experimentation and experiences have not been documented and systematically justified. These practices have been communicated through generations and through words of mouth only.

In PRA, the scientists are subjected to learn the practices from the farmers and try to search/prove logics/justifications including their perception about that. Both these old technologies and the new scientific technologies are required to be blended and not simply replaced. These old practices are called as Indigenous Technical Knowledge (ITK) or Rural People's Knowledge (RPK).

11. Learning Rapidly and Progressively

PRA has been evolved out of RRA, i.e., Rapid Rural Appraisal which was developed with the main objective of getting the results quickly as most of the survey results become obsolate by the time these are compiled and published. Many things change during that period making the results unreliable and unrealistic. Hence, PRA also believes in quick results and speed. PRA result must be obtained in presentation form within a fornight time. The data collection must not last for more than a week.

Further, the data on various aspects are collected one by one and compiled and not all information at a time. The researchers have to be more progressive in respect of items of data to be collected.

Village Entry

Village entry or community entry is the first stage of association of the PRA team with the village community. The team is a stranger to the village community, even if it is equipped with some information of the village. It requires an introduction to the residents of the village before it can start working with them and using subsequent PRA tools. A village entry is undertaken precisely with the objectives of:

* Introducing the team to the village.
* Publicizing the major goal and programme of the PRA.
* Familiarising the village community with the objectives of PRA.
* Ascertaining a positive attitude of the village community to cooperate with the team.

It lands an atmosphere of transparency to the process of interaction between the PRA team and the local community, which aids the process of rapport building. It is said that the first impression is often also the last impression. The time devoted to preparation for village entry can, therefore, be an important factor in the community's acceptance of the team. The attitudinal reorientation is either ignored especially in projects implemented by government departments or not sufficiently addressed. As a result, participatory activities are taken up mechanically without a clear understanding of the spirit or the purpose of the same. The following are some of the pointers for developing the right kinds of attitudes:

* Approach the community with an open mind, with the willingness to change any present notions about rural life and people.
* Have a healthy respect for the rural people and their way - A judgmental attitude will subotage other overtures for eliciting participation.
* Have faith in people's abilities and their intellect. Make an effort to find out how the community have dealt with difficult situations in the past, or how they handle problems. You may be surprised to find the degree of innovativeness that people are capable of.
* Educational qualifications, wealth and social status become the yardsticks for measuring up people, but remember that there is more to life than this. The challenge of life is in living like an "Ordinary" individual and co-existing peacefully with others.
* You are not a provider of solutions for the problems that the villagers face, you are there to learn about the problems and jointly seek a solution to the same. Correct diagnosis can result in quick remedy.
* You are only a visitor to the village, and have certain objectives for interacting with the village community. Do not encourage dependency of the people on you.

Always strive towards building the capacity of the people with whom you interact.

For communities that appear to be responsive, try establishing links with some friendly and/or influential personalities, who may then serve as conduits to the community as a whole. Foremost of all, the team should be aware of its own attitude and behaviour and send out positive verbal as well as non-verbal signals.

Steps of PRA:

- Direct observation by the scientist in the village on village entry.
- Rapport building.
- DIY (Do - it- yourself) helps in rapport building and also gives the practical idea of the complexities involved in doing the activities. Most important technique for rapport building.
- SSI (Semi Structured Interview).

SSI differs from the structured interviewing in its informality. Although no structured schedule or formal questionnaire is used in this kind of interviewing, a flexible outline of the discussion points are kept in mind. As the interview progress, the questions are formulated and asked by the interviewer by keeping in view the central theme of discussion.

Group interviews or focus group discussions are more advantageous than indivudial interview in a sense that they are more participatory and the informations provided are always triangulated. Before conducting any SSI the team members should divide their roles and responsibility during interviewing. To conduct SSI divide the roles amongst team members as —

* Lead facilitator facilitates information flow.
* Process recorder records the proceedings, process and comments.
* Content recorder records information gathered and decisions taken.
* Environment controller - This member of the team controls the gate keepers of information in the village. Gate keeper of information is the villager who is acting as hindrance and who is not allowing other villagers to give the information. Environment controller is from the team of the scientists who controls these gate keepers.

Before SSI (Points to remember)

- The topic of discussion and the specific areas of questions should be determined and discussed among team members before beginning the interview.
- Fix up the time with the villagers in advance

During SSI (Points to remember)

- Greet the group/individual before beginning the interview.
- Set the climate by asking some general questions like weather or any recent incident or some thing which is visible.
- Select a suitable place for comfortable discussion.
- Gradually bring the topic of discussion.
- Always ask open ended questions and try to avoid closed or suggestive questions.
- Avoid irrelevant and very sensitive questions.
- Listen carefully, be patient, try to understand various reasoning on any particular point .
- Do not interrupt or suggest or lecture.

- Always be respectful to villagers.
- Do not ask more than one question at a time.
- No member of team must leave abruptly during the process of interview.
- The interviewer and interviewee should sit in a mixed group in one place and do not polarise the sitting.
- Take a detailed note of all the information.
- Head nodding in agreement or disagreement should be avoided.
- Use local language with local dialect as much as possible.
- Control gate keepers very carefully and allow fence sitters to join discussion.
- Make sure that major information provider are probed and cross checked.
- Record every information as has been said and don't give the meaning or your interpretation.
- Thank people.
- Make arrangements for next meeting. If necessary.

5
Tools of PRA

The practitioners of PRA have developed hundreds of tools for seeking information through the active participation (action) of the villagers depending upon the situations and the type of information to be sought. The villagers can work on these tools by their own hands and minds and thereby their active participation is ensured. No fixed tool can be prescribed for use by the researchers, but they can pick and use according to their applicability under the situations for seeking specific information. Hundreds of tools are available and many more with modifications and still the researchers may be free to choose and even modify the available tool.

This chapter has discussed some of the most popular tools used by the researchers for Participatory Rural Appraisal with their methodology or procedures alongwith preparations required before their use. The tools have been discussed through presenting case examples along with their explanations and interpretations drawn from the earlier studies conducted by the authors during the last decade. Readers are free to make necessary modifications to make more and more applicable and practicable with the intention to seek active participation of as many villagers as possible. The tools have been discussed under space analysis, time analysis and flow analysis.

SPACE ANALYSIS

Space analysis is the study of the agro-eco system in respect of physical space and boundaries, the uses and utilization patterns of physical resources like land, village settlement, crops and cropping pattern, social, economic and occupational patterns. The tools used for space analysis are mapping, transect walk, charts and figures.

Participatory Mapping

"Participatory Mapping" could be a very useful technique for developing participatory land use plan or plan for natural resources management. Outsiders and the project staff can understand the geographical boundaries, roads, water resources, crop fields, houses, wells, schools, buildings and other physical and natural resources of a village easily and quickly from the participatory maps. This helps to understand the villagers level of abstraction because they depend more on the concrete features of the physical environment. Accordingly, outsiders can understand the distribution of the natural and physical resources of the village from the map as perceived by the villagers. Participatory mapping could be specially useful for understanding the topography, forest areas, degraded land, pastures (grasses), water resources etc. This helps in learning quickly the social profile, village enterprises, cropping pattern, land use pattern and the transect of the village too. Mapping includes both map and further presentation of related statistics (distributions) in tabular form or in chart form.

Process:

Before

* Decide what type of map will be drawn and for what purpose.

* Have sufficient discussion amongst the facilitators before starting the process.
* Discuss with the villagers briefly about their land use, low, medium and upland situation etc.
* Try to involve a good number of villagers to prepare the map.
* Colored chalks, colored powder and seeds etc. may be used.

During

* Ask the villagers to draw the map of their own village at a suitable place which is convenient for all the villagers (dry, visible and spacious place, so that many people can look at it easily).
* Allow the villagers to draw the map in their own style indicating geographical boundaries, roads, water resources, wells, houses (men, women, children etc. in social maps) forest, crop land etc.
* Allow them to draw the map in their own way and don't suggest.
* Don't interrupt from outside.
* There could be 5-6 or more numbers of villagers preparing the map of the village at a time. More the participation of villagers better the quality of map.
* Triangulate the information indicated in the map. Ask some of the villagers to point out his/her house, crop field, orchards, etc. on the map.
* Ask others if they would like to add anything in the map.
* Transfer the map on a paper. A legend could be added for better understanding. Write down the names of all the villagers involved.

Types of Mapping

Different types of maps could be used for collecting and analysis of various village resources. Villagers may be asked to draw social map to depict soical stratification in the village settlement, enterprise map to depict the location of various enterprises in the village, land use map to show various uses of village land, etc.

Area under different crops could be presented in a tabular form. Some of the case examples of various types of maps as well as charts with explanations are given below :

Social Map

Social map describes locations of roads and houses, distribution of castes, male female ratio etc. within the village.

The social map in Fig. 1 shows the social profile of Dohna village of Bareilly district of Uttar Pradesh. Dohna village has a mixed type of population with majority of Muslims followed by Yadav caste Hindus. The location of houses of different castes are clearly visible. The village is electrified but majority of the poles seem to be located on the main village lanes. Hand pumps are located at the out skirt of the village.

Enterprise Map

Enterprise map demonstrates the location of houses belonging to various enterprises in the village. The village settlement presented in the enterprise map of Manda village of Bareilly district (Fig. 2) indicates that the households following farming and dairy, farming and poultry and only labour enterprise are concentrated at different locations, while families following mixed enterprises stay together. The enterprises of the village Manda indicate that the majority of the households in the

Fig. 1 : Social Map

Village : Dohna

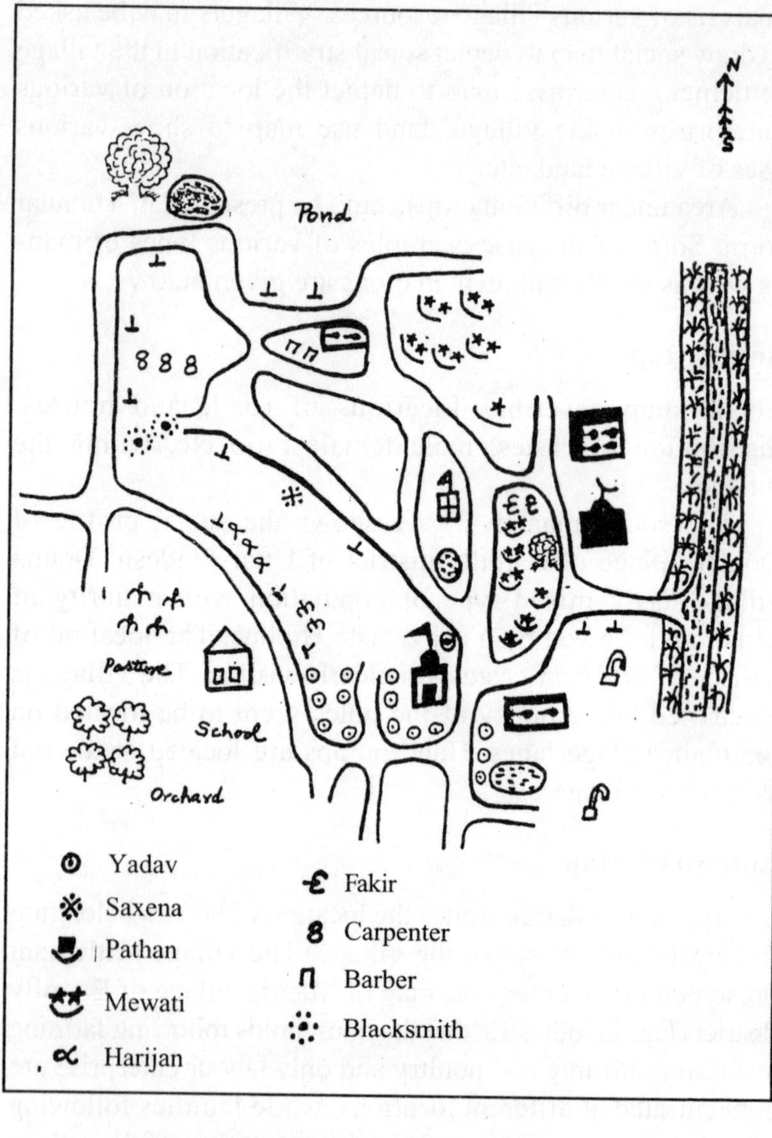

Informers : Farmers' Group

Fig. 2 : Village Enterprise Map

Village : Manda

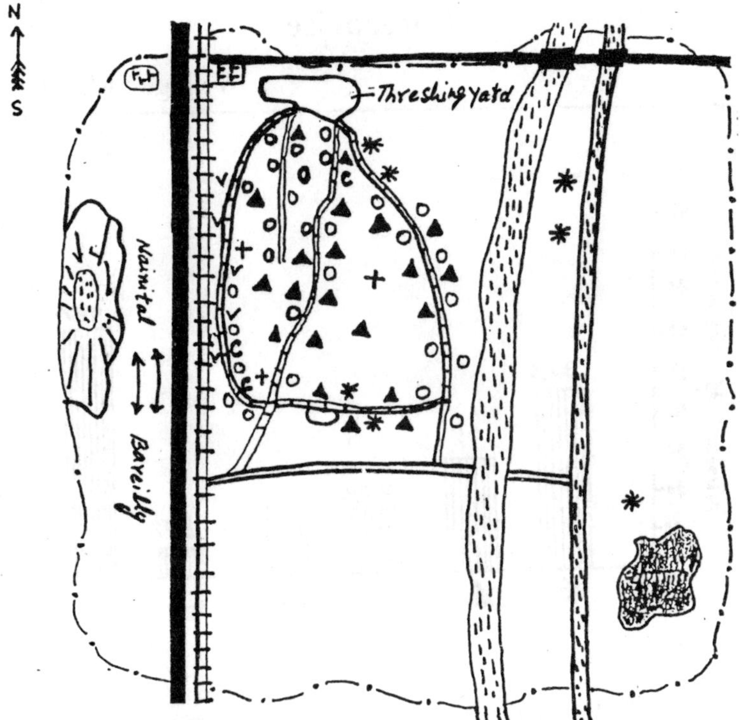

▲ Crop Production

* Crop Production + Dairy

\+ Crop Production + Dairy + Service

v Wages

O Crop Production + Wages

C Crop Production + Poultry

Informers : Farmers' Group

village have farming and labour as their major enterprise (Fig. 3). They possess a small area of land and thus have to earn partly through labour. 32% of the households have a single enterprise of farming, while 7.3% have labour only.

Fig. 3 : Farmers' Categorization According to Enterprise

Village : Manda

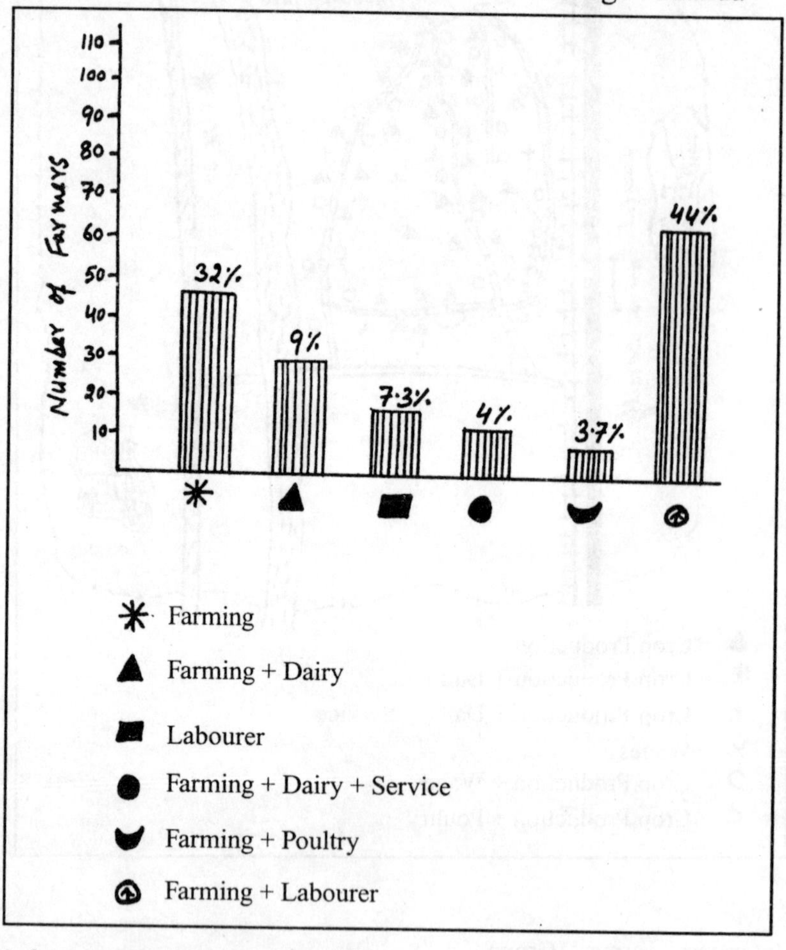

Informers : Farmers' Group

Most of the families (92.7%) have farming as one of the enterprises. While 51.3% have labour as one of the enterprises. 4% of the households also have service as one of the enterprises, while poultry is followed by only 3.7% families.

Land Use Map

Land use map depicts the location / size of different land units according to their use (agricultural fields etc.), location of soil erosions, roads, foot-paths, rivers, public buildings, wells, tanks, pastures, type of soil, orchards, etc.

The land use pattern of Kamua village of Bareilly district, Uttar Pradesh as reported by farmers indicates that village has about 3600 bigha land of which 77% i.e. 2760 bigha is under cultivation. Area under habitation is about 360 bigha while 300 bigha is under orchards. The orchards owned by the villagers are located towards south and east of the village, while sugarcane and paddy are cultivated on north - west side (Fig. 4).

Cropping Pattern

Cropping pattern can best be shown through putting figures in table. The Table-1 depicts the area under different crops of the four villages i.e. Dohna, Manda, Kamua and Purenatal of district Bareilly, Uttar Pradesh.

The total sown area in Dohna village is 5830 bigha. Out of the total sown area, paddy covers the maximum area (2400 bigha), followed by wheat (1800 bigha). Sugarcane is grown in 850 bigha followed by lentil, Berseem, Chari and mustard. About 120 bigha is covered under orchards. Dohna village has the maximum cropping intensity (144%).

Sugarcane is the major crop in Manda village covering an area of 700 bigha followed by paddy and wheat (500 and 425 bigha, respectively). Other crops grown in the village

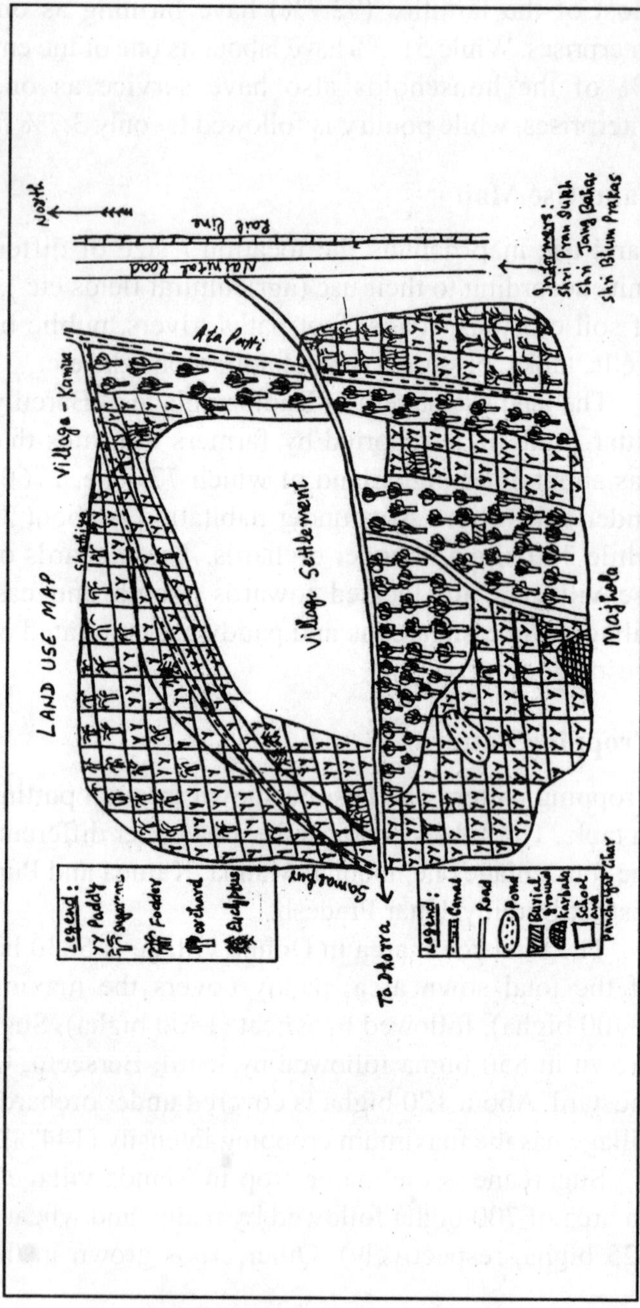

Fig. 4 : Land Use Map

Village : Kamua

Table 1 : Area Under Different Crops (in Bigha)

Crop	Dohna	Manda	Kamua	Purenatal	Total
Sugar Cane/Ratoon	850	700	900	1000	3450
Paddy	2400	500	800	375	4075
Wheat	1800	425	500	425	3150
Mustard	150	175	60	-	385
Lentil	200	75	100	25	475
Gram	30	-	-	125	155
Red gram	20	-	-	-	20
Fodder Crops					
Sorghum	150	100	50	200	500
Berseem	175	60	100	75	410
Horticulture					
Mango	65	30	480	60	635
Guava	55	-	-	1	56
Other Crops	55	25	45	25	150
Total Sown* area (Bigha)	5830	2060	2555	2241	13386
Net sown area (Bigha)	4045	1495	2320	1789	10339
Crop intensity %	144	138	110	125	

* Excluding area under Horticulture crops
Informer : Farmer's group.

land are mustard, Chari, lentil and Berseem (175, 100, 75 and 60 bigha, respectively). Total area under fodder crops is 160 bigha, while mango orchards cover 30 bigha which of course is owned by outsiders.

In Kamua village also the area sown under sugarcane is maximum (900 bigha), followed by paddy and wheat (800 and 500 bigha respectively). A substantial area of Kamua village (480 bigha) is under mango orchards. The other crops grown in the village are fodder crops like Berseem and Chari (150 bigha), lentil and mustard (100 and 60 bigha, respectively). Kamua village has the lowest cropping intensity.

Sugarcane is grown in maximum area in Purenatal village (1000 bigha). Wheat crop covers an area of 425 bigha followed by paddy and fodder crops (375 and 275 bigha, respectively). Area under pulses is 150 bigha. The cropping intensity of the village is 125%

Livestock Population

The information on livestock population can also be best depicted by table. Table - 2 shows that Purenatal village has the highest population of dairy animals, where, there are 125 buffaloes and 85 cattle. Among the cattle population, 60 are cross breds, while Murrah buffaloes are only 15. Almost all the cows are Desi in Manda village except 3 cross breds. Buffaloes especially of Murrah breed are maximum in Kamua village (total 150). Dohna village has 100 buffaloes and 50 cows, almost all Desi except 6 Murrah buffaloes. The village has maximum population of goat i.e. 75, followed by Purenatal, Manda and Kamua. Poultry birds are raised as back yard poultry more in Purenatal village followed by Dohna and Manda.

Table 2 : Livestock Population

	Livestock	Dohna	Manda	Kamua	Purenatal
Buffalo	Murrah	6	10	50	15
	Desi	94	115	100	110
	Total	100	125	150	125
Cattle	C/Bred	—	3	1	60
	Desi	50	75	25	25
	Total	50	78	26	85
Goat		75	25	12	30
Poultry birds		150	20	—	200
Pig		—	—	5	—

Transect - Walk

A transect walk by a PRA team across the village with villagers helps to develop a clear idea on the natural resources, land use, vegetation, changes in the physical features and cropping systems, etc. of a village and share with them a wide range of information and their indigenous features.

A transect walk is a useful method for knowing rural ecological conditions. A walk from one point of the village to another enables the outsider to observe difficult aspects of rural ecology and to discuss with rural people about soil conditions, land use patterns, crops, livestock, micro environments, agro ecological zones etc. Transects help a team to carry out a detailed participatory appraisal of the agro system of a village or an area under consideration.

Process:

Before

- Select group of villagers who have good knowledge on physical resources of the village i.e. tanks, waterhshed, topography, farmland etc.
- Discuss with these villagers the village physical map and mark the route that they would like to follow.
- Distribute responsibilities for recording information among the members of transect group, topic-wise.

During

- Observe and record in minute details of all important things that you come across.
- When talking to villagers feel free to use six helpers i.e. when, what, how, where, why and who to the maximum.
- All team members should cover the enire transect route.

- Collect information about indigenous practices of farmers like soil conservation, irrigation, land use technique etc.
- Request villagers whom you meet on your way to join the walk if they are willing to do so.
- Travel slowly and patiently and try to understand the physical features in the village.

After

- After the completion of the transect walk, sit down at a suitable place with the villagers to have discussion and recording of information or data collected.
- Prepare an illustrative diagram of the transect walk using the information already gathered.
- Generally a transect walk in PRA is done after drawing the physical map of the village.

The transect walk of Manda (Fig. 5) village of district Bareilly, Uttar Pradesh indicate five specific types of lands from west to east. On the south west side of the village there is habitation. On the extreme east, there is pasture land adjacent to which sandy up land soil is located on which oilseeds are grown. The remaining three sects have loam soil which is suitable for paddy and sugarcane. The slopy land with sandy soil has the major problem of soil erosion and uneven distribution of irrigation water. The loamy soil suitable for paddy and sugarcane has the major problem of insects, pests and weeds. Rat is a severe problem in the entire area while termite is visible only in loamy soil. On the north west side sect of the transect is low lying area where paddy and sugarcane is grown.

Fig. 5 : West to East Transect Walk

Village : Manda

Basis for Classification 1	cultivable land 2	cultivable land 3	Land under cultivation and settlement 4	Agril. land with water logged 5
slope of land %	1-3	Normal	2-5	2-4
Soil type	Loam and clay loam	Loam	Sandy and hard soil in South and loam and clay loam in North	Sandy soil at upper side and clay loam at lower side
Land use	Paddy, Sugarcane & social forestry	Paddy and Sugarcane	Land under settlement, Paddy and sugarcane in East; sugarcane, fodder, pulses and oilseeds on upper side in South, paddy, sugarcane, mango orchard, threshing yard in North, Social forestry near village houses	Paddy, sugarcane, chari and gardening and water nuts in pond
Source of irrigation	River water and diesel pumpset	Diesel pumpset	Diesel pumpset and little irrigation by river water	Diesel pumpset and little irrigation by pond
Soil erosion	Light soil erosion	Negligible	High soil erosion	Medium soil erosion
Harmful animals	Rats	Rats	Rats and Termite	Rats

Informer Farmers : 1. Shri Sher Singh 2. Shri Siya Ram 3. Shri Naresh 4. Shri Milkiyat Singh

Fig. 5 (*Contd.*)

Basis for classification 1	Cultivable land 2	Cultivable land 3	Settlement Land under cultivation and Settlement 4	Agril. land with water logged 5
Wild animals	Deer, Antilop	Deer, Antilop Poisonous and water snakes	Antilop, Poisonous and water snakes	Antilop
Problems	→ Reduction in paddy yield due to continuous waterlogging → Lack of alternative crops → Less effect of nitrogen fertilizer → Zinc defficiency → Shelter place of deer and antilop	→ More insects, pests, diseases and weeds → Very low yield → Erosion by river → Damage by deer and antilop	→ High soil erosion → Problem of irrigation and uneven distribution → More insects and weeds → Low knowledge about selection of suitable crop → Damage by antilop	→ Soil erosion → Less water retention capacity → Water logging and uneven distribution → Less alternative crops → Less effect of fertilizers → Zinc deficiency
Possible Solutions	→ Arranging water drainage → Use of organic manure and zinc sulphate → Wild life act may be changed in respect of deer and antilop	→ Study and diagnosis of causes of low yield of crops → Technical information should be provided to check soil erosion by river	→ Agricultural operations perpendicular to slop → Growing of pulse crops → Bunding from place to place → Provide technical information to control insects and weeds → Provide knowledge for selection of crops → Wildlife act may be changed	→ Provide technical knowledge to control soil erosion → Proper selection of crops → Proper arrangement for drainage → Use of zinc sulphate

Informer Farmers : 1. Shri Sher Singh 2. Shri Siya Ram 3. Shri Naresh 4. Shri Milkiyat

TIME ANALYSIS

Time analysis shows the quantitative changes over time and can be used for many variables such as yields, area under cultivation, livestock population, area under fisheries, forestry, horticultural crops, rainfall etc.

For preparing time line, elderly villagers are requested to narrate their life histories. The historical time line provides an historical account of how different aspects of village life have changed over the years.

The historical time line of Manda village, district Bareilly, (U.P.) (Fig. 6) depicts that the village was having only one Pucca house in 1945 and only after 1985 more Pucca houses came into existence in the village. In 1975, three Sikh families came and started improved agriculture in the village and they provided improved seeds to some of the farmers of the village. First diesel and boring pump set was installed in 1979 and in 1985 IVRI adopted the village for livestock development. In 1989 cross breeding and A.I. programme started in the village. Gobargas plant was installed in 1990 and in 1991 first tractor was purchased.

In Fig. 7 the crop time line of Dohna village indicate that before independence the farmers were growing millets in the village. After independence in 1949, first diesel pump set and tractor came in village and farmers started using ammonia as fertilizer in 1950 and in 1954, the villagers started using urea, DAP and Potash. In 1955, the hybrid maize was sown by the farmers. In 1960, the hybrid varieties of wheat were introduced in the village. In 1970, there was severe attack of wilt disease in paddy. The production of Mentha started in 1983 and in 1985 Saryu 52 variety of paddy and HD 2285 of wheat were introduced. The T & V programme started in 1992 while IPM (Integrated Pest Management) started in 1995.

Fig. 6 : Historical Time Line

Village : Manda

Year	Events
1800AD	village settlement
1945	First Pucca house
1955	Use of tiles for ceiling
1965	Installation of Rahat (Persian wheel) and Dhenkali with the assistance of Government
1967	First time a village person selected for Government service
1975	First boring and diesel engine installed.
1979	More Pucca houses constructed
1985	IVRI adopted this village
1986	Kharanja (Pucca village lane) constructed
1989	Livestock breed improvement program started
1990	Gobar gas plant installed and first tractor purchased
1995	village adopted under Gandhi Gram Yojna

Infomation Source : Shri Ahmed Hussain, Shri Sher Singh, Shri Kashmir Singh

Fig. 7 : Crop Time Line

Village : Dohna

Year	Events
1920	Farmers used to grow sorghum, Pearlmillet, small millets like Sawan, Madua, Kodon, Kakani, Leha etc.
1949	People used leafy vegetables of Bathua, Nari Saag and Palak
1950	First diesel engine and first tractor purchased in the village use of Amonium fertilizer started
1954	use of Urea, D.A.P. and Potash fertilizer started
1955	Hybrid maize was sown first time
1965	K-68, Lerma rojo and S.308 varieties of wheat sown in the village
1969	RR-21 wheat varieties was sown in the village
1970	Paddy crop was heavily damaged by wilt disease
1983	Mentha cultivation started
1985	Surya-52 of paddy and H.D-2285 variety of wheat were sown
1992	T&V scheme introduced in village
1995	IPM program started in village

Informer : Shri Kishan Lal, Shri Pratap Singh, Shri Sahab Khan

Animal husbandry time line of Purenatal village, District Bareilly (U.P.) (Fig. 8) shows that the development was started 70 years ago when the first Murrah buffalo of five

Fig. 8: Animal Husbandry Time Line

Village : Purenatal

Year	Events
1920	Shri Umrao Singh of the village first brought Murrah buffalo from Haryana which gave 5 kg milk per day
1945	There were about 300 cattle, 200 buffaloes, 3 horses and one elephant ni the village
1950	H.S. disease spread ni animals
1968	Sorghum and Berseem used to be grown as green fodder
1970	Consultation with veterinary Doctor started
1981	Sahiwal-Haryana crossbred was brought from Rohtak
1985	First dairy was established. Holstein Friesian cow was given to village by IVRI
1990	First Jersey cow was purchased
1994	Dairy was set-up by farmers under Parag Dairy Development Program
1995	About 30 households have set-up dairy by this year

Informers : Sh. Sundar Singh, Sh. Baldev Singh & other farmers litre milk per day was brought from Haryana. The village had been livestock rich since past. Haemorrhagic septicemia disease was spread during 1950. Fodders like Chari and Berseem were introduced during late sixtees and veterinary doctors were consulted for the treatment of animal diseases. The first Jersey cow was brought during 1990 and in 1994, 24 Holstein Friesan cows were purchased. The Parag dairy started dairy development program in the village in 1994.

Seasonal Analysis

Rural livelihood are integrally connected with seasonality. Seasons bring about differences in climatic conditions in crops grown, in availability of water, fuel, fodder, work load, credit demand, incidences of disease attacks in crops as well as in livestocks, which in turn influence the living conditions in rural areas. Seasonal analysis provides the year round activities of villagers on specific aspects. The workload, fodder availability, fuel availability, credit requirements, etc. could be analysed through this PRA tool.

Process:

Before

- Set the environment for the discussion and sit down with villagers in a suitable place.
- Ask them to point out the months in which they have peak of certain things and then the month they have least.

During

- Ask the villagers to display the changes on the floor in any suitable format they like. The data could be indicated using seeds, stones or sticks etc, that is available easily.
- Names of months could be written on small pieces

of paper and be placed in sequential order.

- Ask the villagers to fill the months in which the highest number of activities take place and mark the activity with seeds. Less number of seeds/stones denote a month of less activities.
- Make sure that all the villagers standing around in the periphery get a chance to express their views also.
- Make sure that women participate. Work load analysis should be done by women farmers separately.

After

- Once the seasonal analysis chart has been completed, get other villagers to comment on it and at the same time get it triangulated.

Work Load

The work load in agriculture on men population of Manda village, Bareilly district (U.P.) (Fig. 9) was reported to be maximum in the month of July, while it was high in the months of October - November and then in January to April. The men folk remained almost without work during the months of September and January. The work load in respect of animal husbandry activities was reported to be almost equal in all months except July and September, when it was maximum and in May it was lowest.

Women have high work load in the months of October - November, July and March and very low in the months of December and January. Their work load on animal husbandry activities was maximum in the months of July, October and November.

Fig. 9 : Monthly Workload in Agriculture and Animal Husbandry

Village : Manda

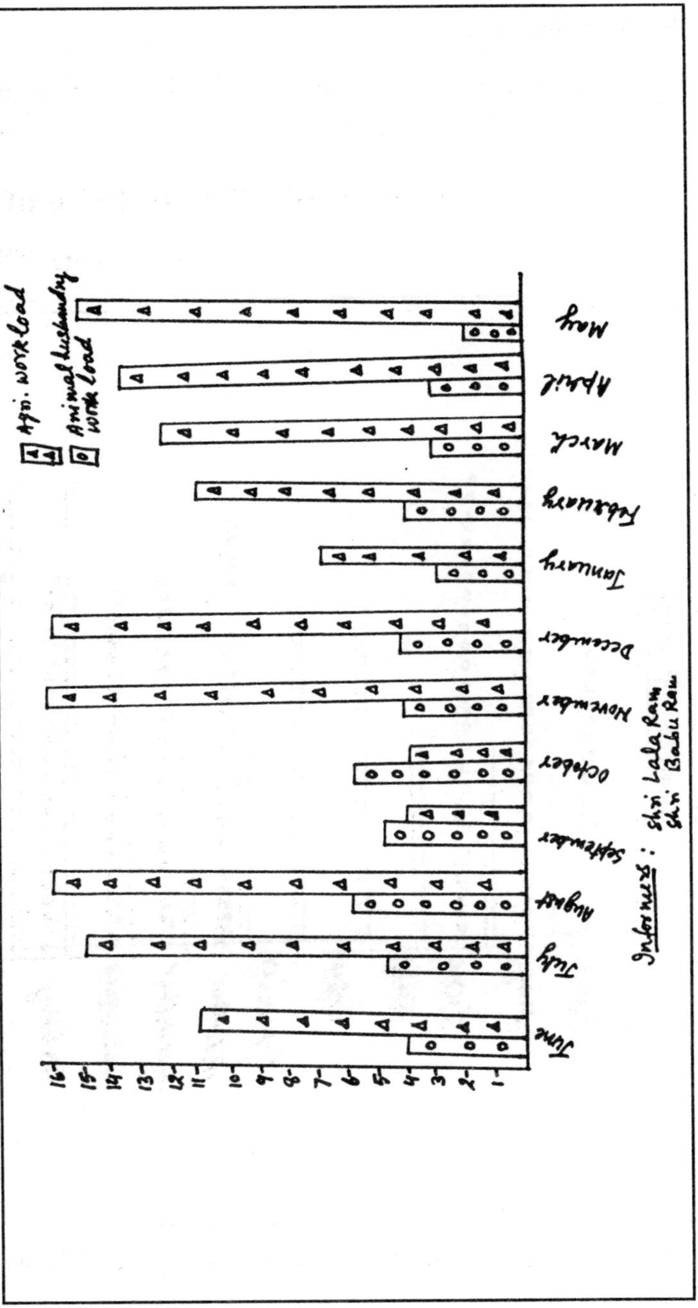

Fuel Availability

The Fig. 10 indicates the information on use and availability of fuel in Dohna village. There is scarcity of fuel in the

Fig. 10 : Seasonal Collection and Use of Fuel

Village : Dohna

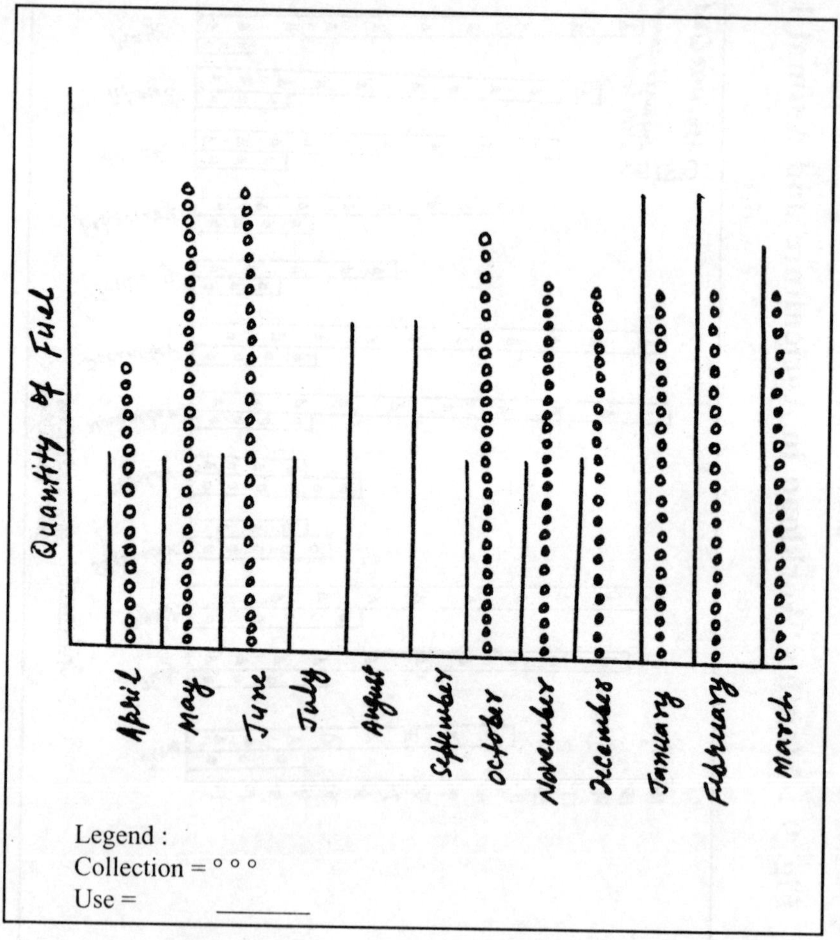

Informers : Smt. Hazara Begum, Smt. Vimla Yadav, Sht. Rajeshwari Yadav, Km. Kumud

months of July - November and hence the stored fuel in the form of dung cakes is used. In the rest of the months the women folk use to make dung cakes / collect wood stick. Generally wood sticks of Arhar and Lehta plants are used as fuels. Sometimes leaves of sugarcane are also used as fuel in the months of January and February.

Fodder Availability

The information on seasonwise availability of green fodder in the selected village Manda of Bareilly district (U.P.) indicated that maximum fodder is available in the months of August, September, October, February, March and April. During the months of February and March, Berseem green fodder is available. In addition to these, green sugarcane top is available from November to March. Green fodder scarcity was reported specially in the months of May and June in Dohna. Scarcity of green fodder was comparatively reported more in Manda village (Fig. 11).

Pests & Diseases in Crops

The attack of insects and pests is more in the months of July, August and September in sugarcane and paddy and from November to March in Rabi crops. Sugarcane crop is severely attacked by stem borer during the months from June to November in Purenatal village of Bareilly district (U.P.). Downey mildew and Mahu problem was reported in mustard crop in Manda village. Rat problem was reported specially in sugarcane and paddy in Purenatal village. Termite problem was reported in Manda village. Smut problem was reported in both the villages. Pulse crops have the problem of wilt disease. Pod borer was reported as the major insect in pea, gram and arhar which has forced the farmers to shift to other crops like sugarcane etc. The attack

Fig. 11 : Month-wise Fodder Availability

Village : Manda

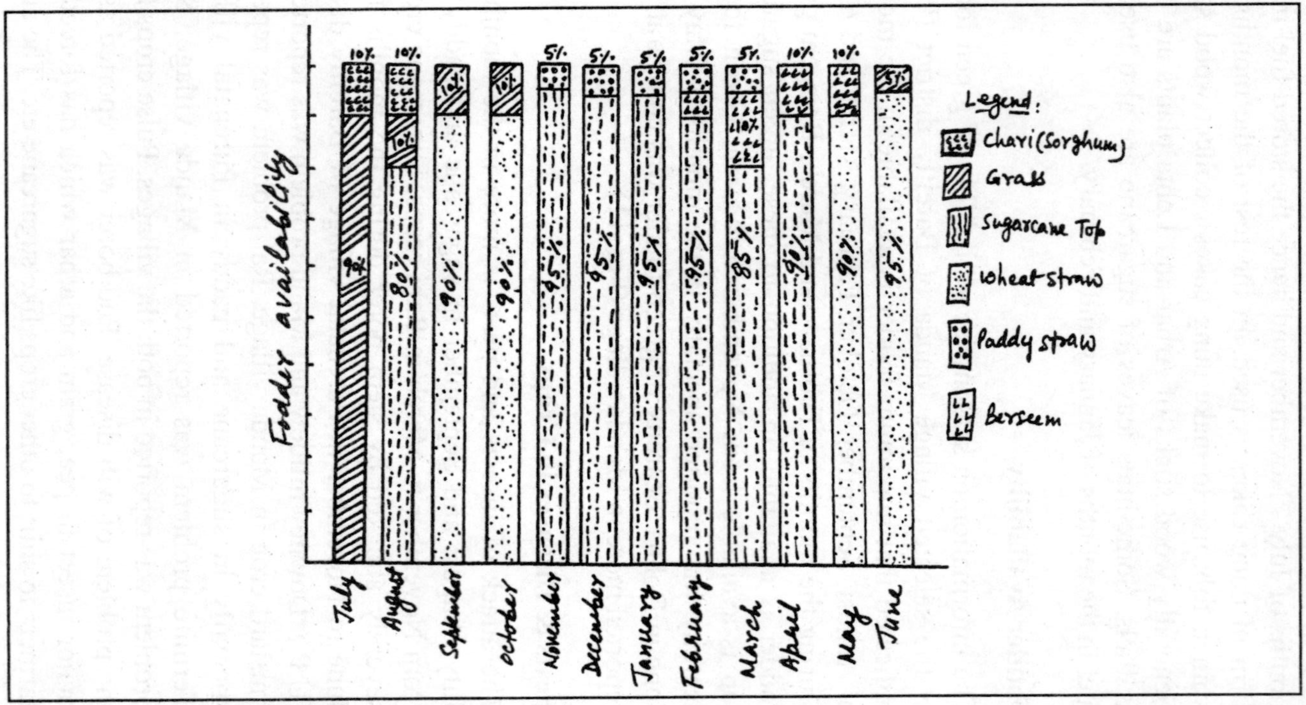

Informers : Sh. Sher Singh, Sh. Bhoop Singh, Sh. Babu Ram

of pod borer is reported during the months of February to April. Wheat crop does not have insect attack but smut and rust problem were reported (Fig. 12 and Fig. 13).

Fig. 12 : Crop Insects and Pests

Village : Manda

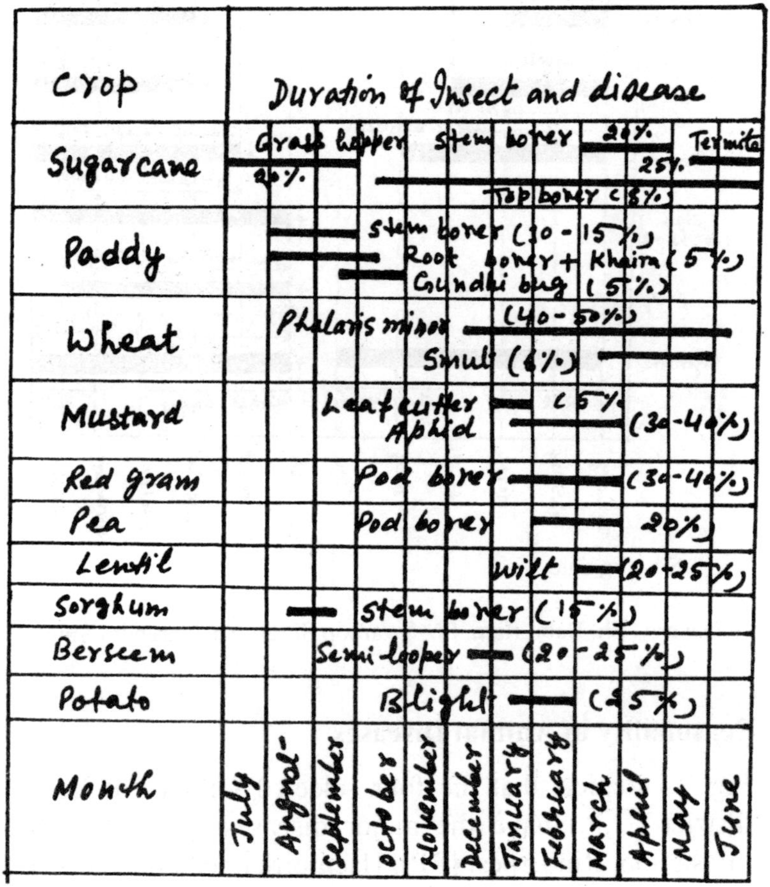

Note : Figures in paranthesis indicate % damage
Informers : Sh. Omkar Singh, Sh. Raja Ram, Sh. Lallu Baksh

Fig. 13 : Duration of Crop Insects and Pests

Village : Purenatal

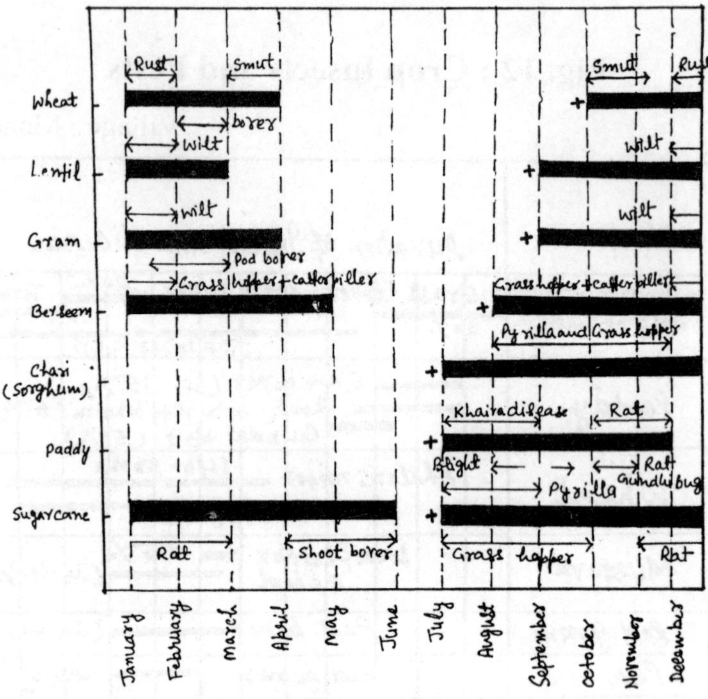

+ Sowing date

Informers : Sh. Babu Ram, Sh. Swaroop Singh, Sh. Narendra Singh

Seasonality of Animal Diseases

Fig. 14 depicts that the four selected villages of Bareilly district (U.P.) have the major diseases/problems of tick infestation, Rinderpest (RP), Haemorrhagic Septicemia (HS) and Foot & Mouth (FMD) disease. The disease periods are almost the same in all the four villages. Tick infestation was found more from July to December. Rinderpest cases

Fig. 14 : Monthwise Disease Prevalence in Livestock

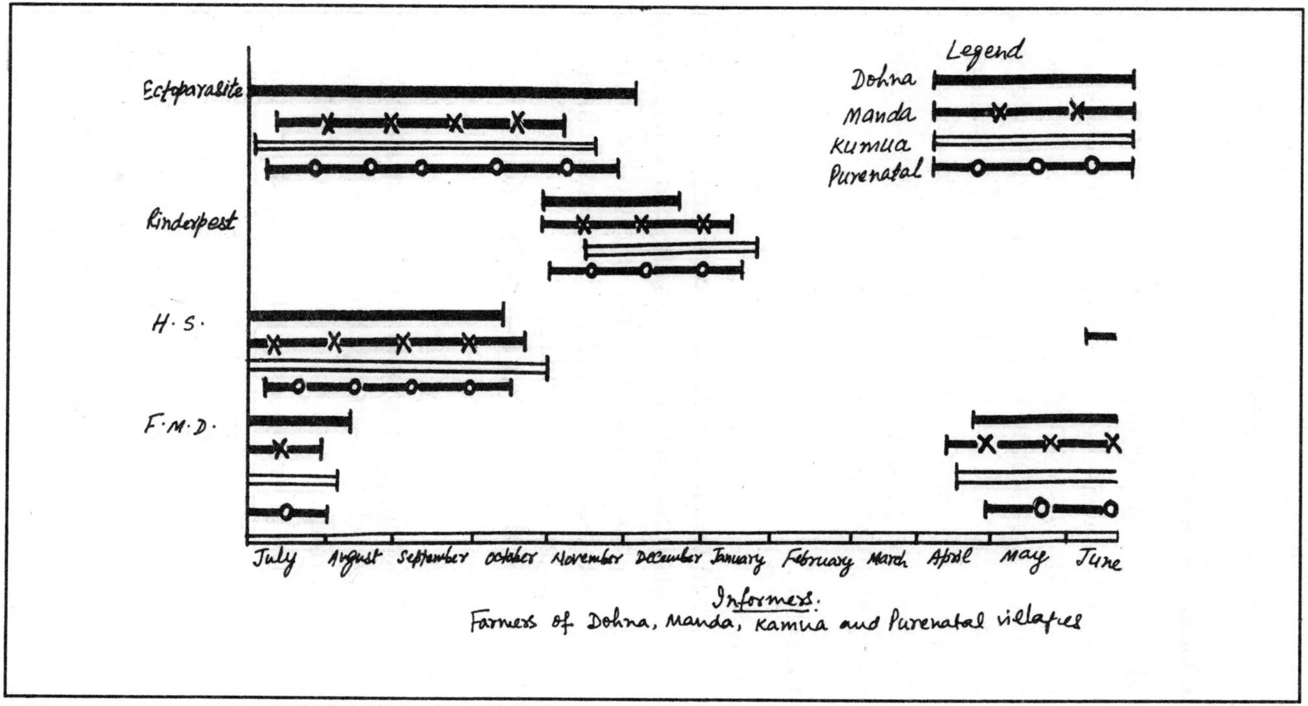

are found during winter season, while HS is spread from July to September. FMD disease is observed from April to July.

Changes and Trends

This PRA tool provides the changes over time in different aspects of village life such as yields, population, livestock population, area under cultivation, use of manure and fertilizer, area under forest, pasture land. etc. Therefore, this technique is useful to the villagers as well as outsiders to appraise some of the development activities expected to be taken up in future in respect to the real change and trends in important areas.

Process:

Before

- Sit down with some elderly villagers who know the historical profile of the village and also some young villagers who have the knowledge in the present socio-economic and environmental changes that are going on in the village.
- Ask the villgers as to how they decide the periods in which the changes and trends are to be analysed.

During

- Start discussion with villagers on a fairly easy topic and ask them to explain how it has changed over the period.
- Then gradually get them to discuss the changes that had taken place in other areas.
- Carefully study the changes and trends in each topic and get the villagers record information on a large sheet of paper or on ground in their own style. Always facilitate them to bring out their own ideas.

After

- When all the information on changes and trends have been collected and recorded, get all other villagers to comment on it so that triangulation of data could easily be done.
- Thank villagers for their participation.

Fig. 15 depicts the changes and trends in Manda village of Bareilly district of U.P. The farmers observed that during

Fig. 15 : Changes and Trends

Village : Manda

Criteria	Year			
	1975	1985	1995	In future
Bullock	XXXXXXX	XXXXX	XX	will reduce
Buffalo Bullock	—	X	XXXX	will increase
Desi cattle	XXXXXXXX	XXXXXX	XXX	will reduce
Crossbred cattle	—	—	XX	will increase
Desi buffalo	XXXXXXX	XXXXX	XXX	will reduce
Murrah buffalo	—	—	XX	will increase
Goat	XXXXX	XXX	XX	will reduce
Poultry	—	—	XXXXXX	will increase
Antilope	—	—	XXXXXX	will increase
Tractor	—	—	XXXXXX	will increase
Improved Implements	XX	XXXX	XXXXXXX	will increase
Improved seeds	—	X	XXXXX	use will increase
Use of fertilizers	—	XX	XXXX	use will increase
Agril. chemicals	—	—	XXXX	use will increase
Green Manure	XXXX	XXXX	XXXX	May increase
Desi Manure	XXX	XXXX	XXXX	May increase
Size of land holding	XXXX	XXX	XX	will reduce
Cropping area	XXXX	XXX	XX	will reduce
Forest area	XXX	XX	—	No possibility of increasing
Pasture	XXX	XX	X	will reduce
Education	X	XX	XXXX	will increase

Informers : Shri Salig Ram, Shri Sher Singh, Shri Prem Singh, Shri Naresh the last 20 years the area under forest and pasture has been brought under cultivation, thereby, increasing the total area under cultivation. However, they observed that in future, there is no hope for further expansion, since no area is now left under forest. With the increase of number of families, the size of land holding has been reducing and will further reduce in future.

Use of agriculture inputs, like improved seeds, fertilizer, insecticides and pesticides, is increasing day by day and will further increase as expected by the farmers. On the other hand, the use of green manure and farm yard manure has decreasing trend. The farmers felt that their use might increase in future. As far as livestock population is concerned, the farmers felt that the number of cross breds and Murrah buffaloes which were not present before 1995 will increase in future and the number of Desi buffaloes which had increasing trend, might decrease in future as in case of Desi cow. Among the draft animals, the number of bullocks has been decreasing, while that of he buffalo is increasing. Goat population increased during 1985 when KVK (IVRI) supplied kids under its lab-to-land programme. The population of the goats however, has decreasing trend. Poultry population would increase further.

FLOW ANALYSIS

With the help of flow analysis, one can observe the patterns of flow of energy, materials, money, information, etc. in a given agro-eco-system. Flow analysis can also demonstrate the cause and effect relationship. The flow analysis can be made with the help of venn diagram, tables, matrix, bar and histograms. Materials and money flow in the rural social system from one place to the other and so also the information. The farmers sell their produce in the nearby market and purchase consumable goods from there. They get the credit from cooperative societies and banks and purchase agricultural inputs from the cooperative stores, government stores or from market. They get treatment from nearby hospitals and get their animals treated at veterinary hospitals. They get farm informations from Block offices, and so on. The farmers may be asked to indicate from where they get such services. The farmers can easily locate these places on the ground at relative distances. Similarly, the farmers can easily compare two or three objects on various critaria with the help of weights on each criteria. By this way matrix can be prepared for further analysis. Flow analysis also include problem cause analysis which the farmers can easily explain with the help of diagram.

Process:

- Inform the group in advance about the meeting.
- Have the objective very clear in your mind.
- Arrange colours, stone pieces, etc.
- Arrive in time.
- State the group the objective and specific purpose of flow analysis, e.g. sale and purchase of animals, comparing the two varieties of a crop, reasons/causes of low yield of paddy crop, etc.
- Select the suitable tool for analysing the specific information like Venn or Chapati diagram, matrix table, flow diagram, etc.
- Ask the villagers to draw the diagram depicting the flow.

Input Purchase from outside Village

The information on fertilizer applicaton indicates that almost all the farmers whether large or small apply fertilizers in sugarcane, wheat and paddy crops. It was also reported that the quantity of fertilizer applied per acre was more in case of small farmers than the large farmers. The fertilizers are purchased by the farmers both on cash and credit either from the cooperative store or from open market.

Fertilizer is one of the main input purchased by the farmers completely from outside village. Among the fertilizers, the quantity of urea purchased is highest followed by DAP and MOP. Per acre consumption of fertilizer was reported to be more in Dohna and Kamua followed by Manda and Purenatal. MOP is used only for wheat crop in Dohna and Kamua villages. Manda farmers apply the highest dose of DAP in wheat crop (120 kg/acre). DAP is not used in paddy crop except in Purenatal village.

Most of the farmers in the four villages purchase seeds of sugarcane, wheat and paddy from within the village and some of them use their own produced seeds. Only few farmers purchase seeds of sugarcane, wheat and paddy from outside the village i.e. from research centres. Number of such farmers is more in Purenatal, followed by Dohna, Kamua and Manda except in case of sugarcane where the number is more in Kamua village after Purenatal.

Availability of Agricultural Implements

The information on availability of different agricultural implements indicate that there is adequate mechanisation of farming in all the four villages (Table-3). Matching to the small size of land holding, all implements viz. tractor, harrow, cultivator, leveller, diesel engine, thresher, dunlop - cart and chaff cutter are available in the villages. The maximum number of tractors are in Kamua village followed by Dohna, Purenatal and Manda. Diesel engine are maximum in Purenatal followed by Manda, Kamua and Dohna.

Table 3 : Availability of Agricultural Implements

Sl. No.	Agricultural Implements	Number			
		Dohna	Manda	Kamua	Purenatal
1.	Tractor	9	2	14	5
2.	Harrow	33	42	25	15
3.	Dunlop Cart	20	12	20	20
4.	Thresher	7	3	15	2
5.	Leveller	2	—	4	1
6.	Cultivator	17	12	7	2
7.	Trolly	5	2	12	3
8.	Chaff Cutter	120	67	90	145
9.	Diesel Engine	20	45	40	65

Infomrers : Farmer's Group

Harrows are maximum in Manda followed by Dohna, while threshers are more in Kamua village. Chaff cutters are maximum in Purenatal village where dairy is one of the major enterprises.

Productivity of Different Crops

The information collected on yields per acre for different crops indicate wide variations among the four selected villages. The per acre yield of sugarcane varied from 150 to 200 qt/acre. The highest yield was recorded in Kamua and Purenatal while lowest was recorded in Dohna village. Similarly, per acre yield of fine paddy was recorded as highest (15 qt/ acre) in Kamua village and the lowest (7.5 qt acre) in Dohna village. The yield of wheat was also highest in Kamua and Purenatal (15 qt/acre) and lowest in Dohna and Manda (9 qt / acre). However, the yield of Lahi was reported as highest in Manda village followed by Kamua and Purenatal.

The net income from mango and guava orchards in

Kamua and Purenatal villages is almost same (Rs. 500-1000 in mango and about Rs. 1200 in guava). The low yield is mainly due to high incidence of pests and diseases (Table-4)

Table 4 : Crop Productivity

(Quintals per acre)

Crop	Dohna	Manda	Kamua	Purenatal
Sugarcane	150	180	200	180-200
Paddy – Fine	7.5	12	15	12
– Coarse	12	—	—	—
Wheat	9	9	15	15
Lahi	—	6	4	4
Mustard	3.6	3.6	4	—
Horticultural Crops (Rs. per bigha per annum)				
a. Mango	—	—	500-1000	500-800
b. Guava	—	—	—	1200

Animal Productivity

The information provided by the villagers (Table-5) indicate that the productivity of buffaloes both Marrah and Desi was highest and same in Purenatal and Dohna villages (8-10 and 5-6 litres /day/ animal, respectively). The productivity of buffalo was 7-8 litres/day/animal for Murrah in Manda and Kamua and for Desi cattle it was 3-4 litre/day/animal in Manda and 5 litre/day/animal in Kamua village.

Table 5 : Livestock Productivity

(litre/animal/day)

	Livestock	Dohna	Manda	Kamua	Purenatal
Buffalo	Murrah	8-10	7-8	7.0	8-10
	Desi	5-6	3-4	5.0	5-6
Cattle	C/Bred	—	8-10	7-8	9-10
	Desi	2-4	3-4	3-5	2-3

The productivity of cross bred cow was maximum and almost same in Purenatal and Manda villages (9-10 litres/ day/animal). On the other hand the productivity of Desi cow was lowest in Purenatal and more in Kamua, Manda and Dohna villages (2-3 and 3-5 litres/day/animal, respectively.

Cost Benefit Analysis of Different Crops

Sugarcane crop was considered to require more funds for its cultivation by the villagers. Manda villagers invest less amount in its cultivation. The production of sugarcane and the net profit was also reported to be highest in sugarcane crop. Next to sugarcane is ratoon crop which provides the highest production and net profit. Ratoon sugarcane crop requires less amount of money in comparison to paddy crop and almost the same as in wheat crop. At this point, this is to be highlighted that sugarcane is two season crop, while others are one season crops.

Basmati paddy was considered as the highest earning crop next to sugarcane in all the villages. The gross income from other paddy varieties is also reported to be very high, but having high cost of production the net profit becomes low. Wheat, lentil, and mustard are next in order in respect of gross income as well as net profit (Fig. 16).

Cost Benefit Analysis of Milk Production

The cost of production as well as the gross production was more for buffalo milk in comparison to cow milk. The net profit was also reported to be more in case of buffalo milk. Purenatal is the only village where cross-breds are maintained. Cross breds top the list of milk producing animals in respect of production and net profit. The maintenance cost of cross breds was also reported to be higher than buffalo and Desi cow (Fig. 16).

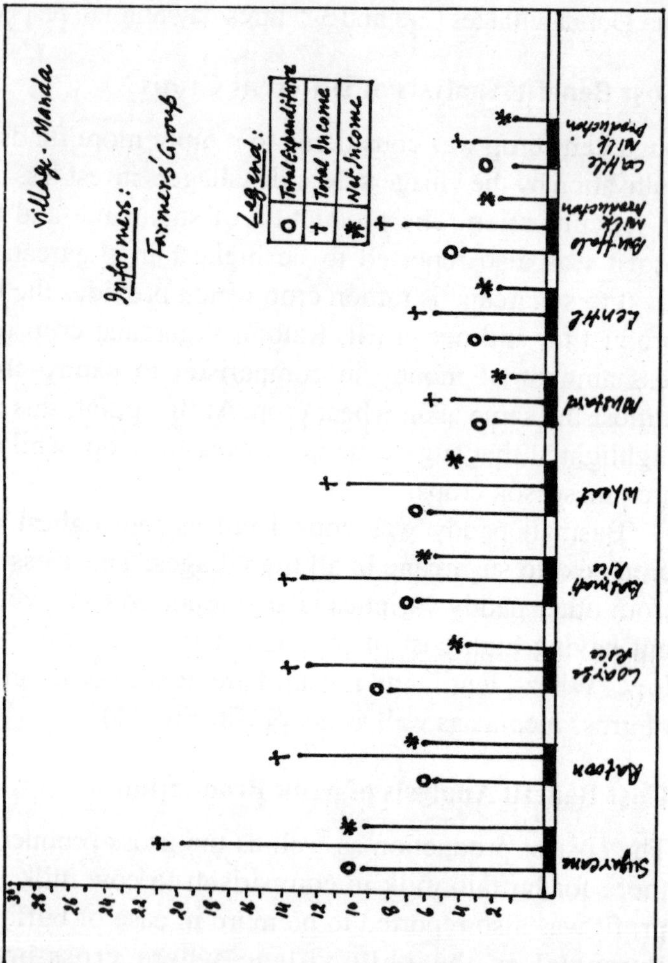

Fig. 16 : Cost of Production, Total Income and Net Income of Various Crops/Enterprise

Matrix Ranking

In this exercise, different attributes and criteria are listed on which alternative practices are ranked either on the basis of fixed scoring (e.g. scoring out of 10) or according to their relative importance. It helps the PRA team to understand the different criteria against which the villagers rank/score any particular item e.g. crop varieties, categories of livestock, source of credit etc. In fact the villagers while they select the best, or the second best or the third, they always judge them against some criteria. This exercise gives an opportunity in understanding the priorities of villagers in selecting any crop, vegetable species, breed of livestock or identifying the more suitable source of credit for them.

Process:

Before

- Decide the topic (s) on which you would like villagers to score.
- Have an initial discussion with the interested villagers, set the climate for discussion and begin.

During

- Ask the villagers to list different alternatives available to them, such as different sources of credit, they have in the village, crop varieties, breeds of livestock etc. Ask them to write these on small cards and arrange them on a horizontal line on the floor.
- Ask them why they prefer one over the other, what are their reasons for selecting. Ask them to write down their criteria for preference on left hand margin one after the other.
- Develop a matrix by drawing horizontal and vertical lines dividing the items written on the top as well as on the left hand margin.

- Start talking about the first criteria. Now ask them to score against only one criteria at a time.
- Ask which one is the best and second best alternative practice and so on and ask them to score them out of 10 accordingly by putting stone or chalk pieces in the respective box.
- Then move on to the next criteria and ask them to score each item written on the top in the same manner as they did earlier.
- Ask the villagers to complete the entire matrix. Do not rush through.

After

- Once the scoring is done, then triangulate it with the other villagers.
- Also ask them their best choice considering all the criteria in the matrix. Thank them all.

Matrix Ranking of Various Crops

Fig. 17 shows the matrix ranking of different crops by the farmers of Dohna village of Bareilly districte (U.P.). The criteria for ranking set by the farmers were definite income, more stability, less cost of cultivation, less pests and diseases, long duration for sowing, net profit and more employment. Matrix rankings were obtained separately for Kharif and Rabi crops.

Among the Kharif crops, sugarcane was ranked as 1st and paddy as 2nd on the basis of more production, definite income, more stability, less pests and diseases and more employment. No other crop was included in matrix ranking under Kharif since the area under other crops was negligible.

Among the Rabi crops sugarcane was ranked 1st followed by lentil, wheat and Lahta.

Fig. 17 : Matrix Ranking For Crop Preferences

Village : Dohna

Criteria \ Crops	Kharif		Rabi			
	Sugarcane	Paddy	Sugarcane	Wheat	Mustard	Lentil
Assured Income	oo	o	oooo	ooo	o	oo
Stability	oo	o	oooo	oo	ooo	ooo
Less cost of production	o	oo	o	oo	ooo	oooo
Pest and disease resistance	oo	o	ooo	oooo	o	oo
Alternative for ploughing and other intercultural operations	oo	o	oooo	ooo	o	oo
Net profit	oo	o	ooo	oo	oooo	oooo
More employment	oo	o	oooo	oo	o	o
Matrix Ranking	I	II	I	III	IV	II

Informers:

Farmers' Group

Matrix Ranking of Paddy Varieties

Paddy is the 2nd major crop after sugarcane in Bareilly district of U.P. The farmers of Manda village (Fig. 18) are aware of several varieties of paddy as Pakistani Basmati,

Fig. 18 : Matrix Ranking of Paddy Varieties

Village : Manda

Criteria \ Varieties	Pakistani Basmati	Russi Basmati	Indralam	Pant-4	Saket-4	7112
High yield	••	•••	•••••	••••••	••••	••••
Less fertilizer requirement	••••••	•••••	•••	•	••••	••
Insect Kesistent	••••••	•••••	••	•	••••	••••
Disease Kesistent	••••••	•••••	••	•	••••	•••
Early maturity	•••	••••••	••	•	•••••	••••
More keeping quality	••••••	•••••	••••	••	••	•••••
Less cost of production	••••••	•••••	•••	•••	••••	••••
Net profit	•••••	••••	••	••	•••	••••••
Matrix Ranking of farmers	I	II	VI	IV	V	III

Informer farmers: Shri Baljiar Singh, Shri Sher Singh, Shri Om Prakash

Roosi Basmati, Indrasan, Pant 4, Saket 4 and Pant 7112. Among these, Basmati varieties were ranked as Ist and 2nd respectively, though having the lowest per acre yield. The high ranking was given due to its best taste, less requirements of fertilizers and less insect/pest and disease attack. Pant 4 was mentioned as the highest yielder but the poorest in other criteria, especially in respect of high attack of pests and diseases.

Matrix Ranking of Fuel

Purenatal village women of Bareilly district (U.P.) especially the scheduled castes were very much particular to indicate the matrix ranking of different fuels used for cooking (Fig. 19). The women reported eight different sources of fuel viz. cow dung cake, dry leaves of sugarcane, Dhencha sticks, Pakad wood. Cow dung cake was ranked as 1st for its availability throughout the year, less smoke and more fire. Dhencha sticks were ranked as 2nd due to their easy availabiltiy, less smoke and more fire. Though, Pakad wood is the best reported one in respect of less smoke and more fire but is not easily available, hence ranked as 3rd. Dry leaves of sugarcane are the most easily available fuel and gives less smoke but produces very little fire/heat. The matrix ranking of fuel indicates very clearly that the farm women prefer cow dung cake in comparison to any other fuel available. Under the situations, the other alternatives will have to be thought of for fuel requirement if cow dung is to be saved for FYM.

Matrix Ranking of Livestock

Fig. 20 shows that in Kamua village of Bareilly district (U.P.) different breeds of cattle and buffalo ranked separately. Among the cattle breeds, cross bred was ranked as 1st

Fig. 19 : Matrix Ranking of Fuel

Village : Purenatal

Qualities \ Type of Fuel	Cow dung Cake	Sugar Cane Top	Pakad Tree wood	Sugar Cane Roots	Dhaincha wood	Sun hemp Stem	Arhar Stem	Mustard Stem
Less time in collection	3	8	2	1	7	5	4	6
Easy available	4	8	1	4	7	6	6	6
Don't blaken utensils	7	1	8	2	4	2	1	2
Available whole year	8	4	7	4	4	4	4	4
More flame	7	1	8	5	6	2	2	2
Less smoke	8	7	8	7	7	7	1	7
Ranking	I	IV	III	VII	II	VI	VIII	V

Informer farmers :

Smt. JagDevi

Smt. Ramvati

Smt. Shanti

Smt. Sushila

Smt. Lilavati

Fig. 20 : Matrix Ranking of Cattle and Buffalo Breeds

Village : Kamua

Attributes	Cattle			Buffalo	
	Desi	Haryana	Crossbred	Desi	Murrah
More per day milk yield	1	3	4	4	7
More lactation period	2	3	4	4	7
Disease resistance	4	3	2	7	6
Less cost in rearing	4	3	2	7	5
More sale price of adult calves	2	4	4	4	7
More income	1	3	4	5	7
Matrix Ranking	III	II	I	II	I

Informers :

Farmers' Group

followd by Haryana and local (non-discript). Cross bred was considered to be the best in respect of its high production, long lactation period, better price of calves and more income. However, the cross breds were reported to have more disease prevallence and require more cost of rearing. Desi cow was reported to possess more disease resistance.

Among Desi and Murrah buffaloes, Murrah was ranked as Ist for the same criteria as in the case of cross bred cattle.

Quantifying Indigenous Knowledge (QuIk Method)

Indigenous Technical Knowledge (ITK) is generated by local people through a series of innumerable informal experiments. ITK is a valuable national resource; its systems are dynamic and never static. It is cost effective since it builds on local resources for enhencing productivity and promotion of sustainability through rationalization of short/long term goals.

Validation of ITK is a logical step so that the effectiveness of the practices can be quantified. Modifications of the practices, if required will help to develop a build on that will be appropriate and acceptable for the farming situations. A reliable method has been developed for evaluating and validating ITK through PRA matrix ranking. This is called Quantifying Indigenous Knowledge (QuIk) method.

QuIk method for quantification of Indigenous Knowledge was developed by Villiers (1996). Which is based on the assumption that farmers know and understand the conditions in a better way in which they work. In QuIk, the matrix ranking, one of the PRA tools is combined with a semi-structured interview to find-out the numerical data from experienced farmers / farm women with the assumption that their responses are simulation of the results of their informal experiments.

QuIk develops the PRA tool matrix ranking/rating to access farmers' technical knowledge in order to test hypothesis about agricultural performance. Criteria for the matrix are defined in order to test a specific hypothesis which may itself be derived from interactive farmer/ researcher discussions. For example, farmers may have made the general statement that disease incidence in turmeric is higher in high fertility shaded sites. The matrix is then designed to test this hypothesis in a similar way that a randomized block field trial would be designed. That is, the factors and their levels are set in such a way that the null hypothesis (that all treatments will give the same results) can be tested. For the turmeric example the following treatments may he selected : shade and high fertility, shade and low fertility; no shade and high fetility and no shade and low fertility.

After field testing the matrix is used as the basis for the interviews with a series of independent key informants. Experienced farmers are identified who have been using the technology, through informal inquiry in the village. A method described by Ashby *et al.* (1989) gives a systematic process for identifying the reputed persons (defined by years of experience and interest in cultivating a crop). A series of informal interviews on the wider subject are carried out. Each farmer who is interviewed is asked to give his / her opinion as to who are the local experts in the particular crop. The additional farmers identified by this method are visited, with the result that the list of names is expanded and certain individuals are repeatedly mentioned. Those farmers whose names occur more than twice may be considered experienced and appointments can be made to invite them to join the more focussed activity.

The matrix is designed through preliminary discussions

with farmers and then is used as part of a systematic process to obtain quantitative data. The same matrix is used to interview a number of farmers and the data from each farmer is treated as an independent result. The data set from all farmers can then subjected to statistical analysis through ANOVA. In this way QuIk is used to test the hypotheses about crop/livestock performance in a simulated trial, very similar to the traditional field trials used in agricultural research. The effects (and their significant differences) of theoretical experimental treatments on crop/animal responses may thus be evaluated without costly field trials.

QuIk represents one way of conducting these interactions. QuIk data is drawn from holistic perspective desired from farmer assessment within his/her livelihood system. Through combining the data from a number of farmers, and subsequent statistical analysis, the interpretation of data becomes broadly applicable but remains relevant to the individual farmers (Villiers, 1996).

Steps for Qulk Method:

- Conduct informal enquiry in village to identify experienced farmers.
- Contact these experienced farmers and ask them to give their/opinion as to who are the local experts in the particular crop/animal husbandry.
- The additional farmers identified by this approach are visited.
- The list is expanded and certain individuals are repeatedly mentioned.
- Farmers whose names appear more than twice are considered experienced and selected as key informants.

Matrix Ranking Process:

- The matrix is designed in the form of a randomized block field trial.
- Draw the matrix table on the ground.
- Label the columns and rows with criteria and treatments
- Carefully explain the treatments to the farmers.
- Ensure that all farmers refer to the same parameters.
- Stones or other locally available materials can be used for scoring.
- Ask to allocate the stones to various cells in matrix.
- The data from each farmer is treated as an independent result.
- The same matrix is used to interview a number of farmers.
- The data is subjected to ANOVA for statistical analysis

Merits of QuIk:

- Enables agronomic/livestock parameters to be quantifed without field trials.
- Alternative to field trials.
- Far lower cost than field trials and less time consuming.
- Quantification of qualitative data can be done
- Particularly useful for long duration crops and for livestock

6
Participatory Research and Participatory Technology Development

The great strides in agricultural growth have not only made India self-sufficient in food but have helped also in creating highest ever buffer stock (63 million tonnes) as well as considerable resilience to the system to the weather vagaries and natural calamities. However, burgeoning demographic pressure of both human beings and livestock is continuously exerting demand for food and other agricultural products on our already over-exploited natural resources. These productivity gains are by and large at the cost of environmental degradation and other serious concerns for agriculture. The growth has not benefitted the farmers of lesser endowed areas, particularly rainfed agriculture who do not have enough capital resources- widening the disparities in overall economic growth in the country. Unfortunately the gulf between 'haves' and 'have nots' is increasing day-by-day leading gradually to social upheavals.

The Rao Committee (1993) has rightly pointed out that breakthrough in agricultural production scenario is successful only in the core areas, which are blessed economically, politically and environmentally, as is evident from the fact that more than 70 percent of the food production is contributed by only 56 districts out of above 500 districts in the country. Even if the researchers and extensionists could transfer and demonstrate the technologies, a majority of them often go unused once external support is withdrawn. On the other hand if the technologies are sound and viable, it will become a part of the integrated production technologies even without the intervention of external agencies.

The failure of green revolution to have much impact in the more extensively framed peripheral environments suggests a failure to comprehend the needs of those environments which are often heterogenous, economically weak, risk prone and complex diverse farming systems. Several studies have proved that the technologies developed on research stations are not adapted to the complex, diverse, resource poor and risk prone environments prevalent in farming situations. The 'on-station technology packages' do not suit the varying micro farming situations in an agro-ecological zone, unless they are refined and fine-tuned to suit the recommendation domain. It is very much relevant in India where farming is done in diverse environments, viz., dry farming or wet farming, coastal system or hill region system, upland system or low land system, sandy or loam or clay soil systems, etc. Even within one agro-eco system itself, the needs of diverse farming situations have to be addressed which may not be possible by conventional on-station-research.

The conventional research system gives priorities of researchers, their needs and interests and outcome may or may not suit for the target group. Further, the integraton of the three systems involved in technology development and adoption, viz. research, extension system and the farmer, is practically very poor. Hilderbrand and Poey (1985) said, "the conventional research system gives an estimate of what would happen if farmers were to control the variables as the researcher does. It does not furnish an estimate of results, if farmers were actually to use the new technology."

Recently, world wide research approach is shifting towards working with the farmers and for the farmers. Producer friendly technologies are likely to be evolved through reorganizing the structure and process of the technology development system. All the three systems in technology development, transfer and adoption (research, extension and farmer systems) will be able to articulate their perception of problems provided that all these systems are involved in the various processes from identification of the problem to the solution of the problem. The Johl Committee of ICAR (1995) in its report emphasized the need for developing appropriate research programmes through farmer participation in order to account for complexities of required knowledge involved and the ecology as well as the social environment in which the farmers work.

The lack of close working relationship between national agricultural research system (NARS) and extension organizations (first between the research scientists and between extension personnel themselves, and then between and among the two broad categories) and with ultimate users including farmers and agro-industry managers is one of the most complex institutional problem with very high diversities.

Different agencies/countries have recognized various constraints in linkage development and several advances in this direction have been made to remedy them. Linkage mechanism refers to the arrangements, procedures and channels which are used to strengthen organic links in the technology generation, technology transfer and adoption processes. The mechanisms followed under different conditions by various agencies are : linear model (Research - Extension - Farmer), Farming Systems Research and Extension (FSR/E), Training and Visit system, five element (MOTRM) model, knowledge systems and convergence models, participatory technology development model, chain-linked model, flow system analysis model and the first line extension model. Still no individual or agency has come forward with an universal institutional model that could be effective in different agro-climatic, biological, socio-cultural and economic conditions.

The role of scientists and extensionists do not end with technology generation and transfer only but with technology adaptation, adoption and feedback. The role of farmer participation is more in adaptive and applied research and very limited in basic research. In view of this the basic and strategic research are best done on the research stations or under controlled conditions, while on-farm participatory research is more appropriate to applied and adaptive phases, where technology is adjusted to the specific needs under a particular set of conditions.

Another area to concentrate is the validation and integration of Indigenous Technical Knowledge (ITK) in the farm production systems. It is to the credit of these ITK systems that village communities have been able to successfully adjust to the emerging situations, changes and challenges. Farmers with no formal scientific training are always engaged in informal adaptive research and to ignore this valuable knowledge system is a serious flaw in the formal research system. Rural communities have a reservoir of expertise in management of complexities in agro-eco systems. The challenge before the scientists is to properly document the ITKs and integrate them into the production technologies after validation. This would be prossible by research in participation with the extension systems and the rural farmers.

Several terms (ORP, OFR, OFT, FPR, PTD etc.) related to participatory research are in existence, which may be having different concepts and methodologies. Participatory research and technology development requires involvement of scientists, extensionists and the farmers. Participatory Technology Development (PTD) is "a process in which the knowledge and research capacities of farmers are joined with those of scientific institutions, while at the same time strengthening local capacities to experiment and innovate." Farmers are encouraged to generate and evaluate indigenous technologies and to choose and adapt external ones on the basis of their own knowledge and value systems. Although this means that technology development must involve farmers, it does not mean that scientific research has no place. Research will have to contribute on many fronts, such as in the development of resistant cultivars, knowledge of life cycle of pests, process in nitrogen fixation, etc. But, all these contributions must be seen as providing choices for farmers in making farm-specific decisions.

Number of agricultural development projects have demonstrated that agricultural production can be improved in resource poor regions also through the adoption of technologies, provided that farming households are fully involved in the generation of technologies, in their extension to other farmers and in the experimental adaptation to local conditions. In many locations, small revolutions in institutional support and farming practice are occuring; returns to investment are rapid; the poor became better off and more secure and the region produces surpluses.

Researchers and farmers participate in different ways depending on the degree of control of each factor has over the research process. The most common form of participatory research is researcher designed and implemented, even though it might be conducted on the farmer's fields. Many on farm trials and demonstrations represent nothing better than passive participation.

Typology of Participation

Pretty (1994) described the following seven grades of farmers' participation in technology development and transfer.

1. *Passive participation* : People participate as listener or by being told what is going to happen or has already happened. It is unilateral announcement by the scientist without any listening to people's responses. The information flow is down ward and being shared belongs to the external professionals.

2. *Participation in information giving*: People participate by answering questions posed by scientists using interview schedule or similar approaches. People do not have opportunity to influence proceedings as the findings of the scientists are neither shared nor checked for accuracy.

3. *Participation by consultation* : People participate by being consulted and scientists listen to their views. Scientists define both problems and solutions and modify these in the light of people's responses. Such a consultative process does not concede any share in decision making and scientists are under no obligation to take on board farmer's views.

4. *Participation by material incentives*: People participate by providing resources, viz. land, water, labour, etc. in return for cash or other material incentives. Much on-farm-research falls under this category, as the farmers provide the fields but are not involved in the experimentation. It is very common to see this so called participation, where people have no stake in prolonging activities when the incentives end.

5. *Functional participation* : People participate by forming groups to meet predetermined objectives related to the mandate of the research institution. Such involvement does not tend to be at the early stages of project cycle or at planning, but rather after major decisions have been made by the scientists.

6. *Interactive participation*: People participate in joint analysis, which leads to action plans and formation of new local institutions or the strengthening of existing ones. It tends to involve interdisciplinary methodologies that seek multiple perspectives and make use of systematic and structured learning processes. These groups take control over local decisions and so people have a stake in maintaining structures or practices.

7. *Self-mobilization*: People participate by taking initiative independent of research institutions to change systems. They develop contact with research institutions for resources and technical advice they need, but retain control over how resources are used.

Type of Farmer Participation in Research (Biggs, 1989)

1. Contractual - Scientists contract with farmers to provide land and other services like water, labour, etc.
2. Consultative - Scientists consult farmers about their problems and then develop solutions.
3. Collaborative - Scientists and farmers collaborate as partners in the research process.
4. Collegial - Scientists work to strengthen farmers' informal research and development system.

PTD in agriculture is essentially the process of purposeful and creative interaction between the rural people and scientific communities. Through this interaction both the partners try to increase their understanding of the main traits and dynamics of local farming system, to define the prioritized problems and opportunities and to experiment with the selection of "best bet" options for improvement. The options are a synergy of ideas and experiences derived from the indigenous knowledge and frontier science (both from local and from farmers elsewhere). This process of development is geared not only towards finding solutions to current problems, but also towards developing sustainable agricultural practices which conserve and enhance the natural resources so that they can be used for future generations. Most important part of the PTD is to strengthen the capacity of farmers to analyse their pressing problems and to develop relevent feasible and useful innovations at farm level.

Participatory Technology Development is a process in which :

* The problems for which solutions have been sought are defined by farmers.
* The alternative solutions to be tested are defined by farmers.
* The design of experiment is decided by farmers.
* The implementation is done by farmers.
* Monitoring is decided and executed by farmers.
* Evaluation is done by farmers.

Features of PTD

* Focuses on farmer led experimentation.
* Relevent problem solving measures are tested.
* Local knowledge and resource base of farmers are considered.
* Locally available resources are made use of.
* Sustainable productivity is focused.

Steps in PTD Process

Step	Activity
1. Getting started in the field	* Building relationship/rapport * Preliminary situation analysis * Awareness mobilization
2. Understanding problems and potentials	* Analysing driving/restraining forces * Identifying scientific and local knowledge
3. Looking for things to try	* Identifying priorities * Screening options, choosing selection criteria
4. Designing experiments	* Developing agreed research agenda * Review existing practices * Planning and designing experiments
5. Trying Out	* Designing evaluation protocols * Implementation of farmer experiments * Measurements/observations of research data * Developing structures for dissemination of experiences. * Analysing the results
6. Evaluation of the experiments	* Farmer extension - conduct of field days / exchange * Communication of principles and result of PTD process
7. Sharing results with others	* Training in skills and use of experimental methods * Creation of favourable conditions for continuing experiments
8. Sustaining the PTD Process	* Developing self management capacity of farmers * Farmer consolidation * Gradual withdrawal of the organization
9. Scaling up/phasing out	* Dissemination to more farmers

Advantages of PTD

* Farmers produce their own technologies:
  - appropriate to farm situations,
  - based on problems faced by farmers.
* Encourage community participation.
* Use local materials and local expertise.
* Cheap and flexible.
* Culturally supportive.
* Reduce time leg in large scale adoption.
* More sustainable and productive.
* Within the resources and capabilities of the farmers.
* Option rather than fixed packages.

7
Participatory Research For Women Empowerment

Women are considered as second grade citizen in the rural areas due to prevalent male dominated socio-cultural values and norms of the society. They are supposed to bear the entire burden of household work and the bulk of economic activities in agriculture specially in animal husbandry and dairying. Still they have very little say in economic decisions of the family. They are not given any monetory share or other wise for their hard work while the men folk control the expenditure pattern of the family and can invest separately on themselves. As such the rural women can not invest money on themselves and are therefore always economically poor in practical terms.

The major cause of poor economic condition of women is that they do not have suitable independent enterprise through which they can earn separately. On the other hand they are not fully employed and have adequate leisure time, but, they lack in knowledge and skills. More than this they have no opportunity and inspiration for their own development. Participatory research and development could be the best method for their own understanding, inspiration, experimentation and development of knowledge and skills for their empowerment.

Following is the case where, participatory research was designed for women empowerment under Pilot Project on Technology Assessmnet and Refinement through Institution - Village Linkage Programme (IVLP). Two enterprise modules were considered appropriate to be tested on the farm women in the selected four villages of Bareilly district. These were (i) backyard poultry and (ii) goat rearing of Barbari breed. Farm women can independently maintain poultry birds and goat and can earn separately. 20 poultry chicks of 15 days duration or two goat kids were placed with one farm women. They were supported with regular technical advice. Critical inputs were also arranged.

The following case of backyard poultry provides a useful experiences and learnings on women empowerment through targetted participatory research and develpment.

Improved Breed of Poultry under Backyard System

The module was introduced with the objective of testing the adaptability of improved poultry strain, increasing income of farm women and for providing economic security to farm women. One hundred farm women were selected from four villages who needed economic support and security and who were willing to take up poultry enterprise.

Twenty cari-bro-91 strain broilar chickes were placed with each farm woman under backyard poultry system. The farm women were advised to construct small katcha poultry house as per their prevalent practice and material availability. These were made up with mud with one small window of wooden rods. On some of the houses they used wire mesh in place of wooden rods for ventilation. The houses were very small in size where poultry chicks hardly sit and cannot move. The chicks were placed under backyard system with almost zero input. Usually, these were let free during day time where they used to feed themselves with grains, food stuffs and insects from kitchen waste and other agricultural wastes thrown out by the farm women. To supplement, some of the farm women used to feed little food grain, paddy or rice two to three times a day.

Results

Weight gain in poultry chicks was found to be satisfactory in almost all the units under zero input conditions of backyard system. This ranged from 800 gm to 1 kg. within 60 days. Weight gain was more where more space was provided to chicks in poultry houses. The enterprise faced a set back during the first month for the outbreak of fowl pox disease which caused high mortality. Overall mortality was, however, remained at 38.70%. In addition to fowl pox, there were certain accidental mortality also. It is necessary to see and analyse the actual cause of outbreak and severity of fowlpox disease for which further experimentation is suggested. In certain cases, the whole flock of 20 chicks was died. The mortality was heavy where the chicks were placed in insufficient space.

In all 1,075 broilers were sold worth Rs. 38,147.00 even after heavy mortality as was reported by the farm women. There has been a general tendancy among the rural people to hide the actual income. The price of a broiler was told to be only Rs.35.50 where as certain farm women could indicate that a broiler was sold at Rs. 70.00 to 80.00. On an average a farm woman could earn more than Rs. 400.00 within 2 ½ months. Per broiler profit was more than Rs. 30.00 under such adverse conditions. The scientists estimate of Rs. 15.00 per broiler profit was not only proved but it was more than double if mortality is not counted.

The poultry enterprise not only provided an additional income directly to farm women but also served as an economic security to them. The only risk in this enterprise under rural conditions is to ensure against epidemic diseases to reduce the mortality. Because of very heavy mortality and at the same time very high potentialities, further testing of poultry enterprise under the prevalent situation was suggested.

Through little motivation by project scientists, this participatory research and development with 'backyard poultry' module helped in empowering the women. This became a unique enterprise where women have the full say in decision making right from the production to the sale and in making use of its earning. Males did not have any say. This was possible only through participatory development process with farm women, the target group.

8
Participatory Planning

The process of programme development is fundamental to the work of an organization. A well thought plan is one of the requirement for undertaking any activity to achieve successful results. Planning in advance keeps the wheels rolling in the direction of established objectives and goals of a programme. The planning is defined as a process made up of sequential decisions and actions to be taken. Planning is a way to anticipate and offset change. It forces the programmer to look ahead and anticipate future deviations based on data from the past and present, to consider the impact of changes on the programme and to develop appropriate responces. Thus, planning is an information handling process or system.

Careful planning is necessary for the success of any developgment programme. It is necessary to concentrate on goals and objectives, to make the most effective use of available resources of land, labour, human, material and time to provide the continuity of efforts and accomplishment of objectives and for timely utilizing the resources to achieve both the short term and long term objectives.

Programme planning encompasses a series of steps According to Pesson (1966) Programme development process is a continuous and cooperative activity of people and development workers in which problems are identified, objectives are set and actions are taken to reach the objectives. Programme planning involves eight steps, viz. collection of facts, analysis of situation, identification of problems, decisions on objectives, development of plan of work, execution of plans, determination of progress and reconsideration.

Effectiveness of a development programme has direct relationship with the involvement of the prople for whom the programme is meant. Moreover, India has adopted the goal of development through democratic approach. Under such situations people's participation is basic to the success of development programes. It means participation of all beneficiary people irrespective of caste and creed, rural or urban, men or women, young or old, educated or uneducated.

The participation in planning by intended beneficiaries may yield the following advantages :

1. During designing phase, the beneficiaries can provide additional information on cultural, ecological, economic and indigenous technical knowledge.
2. Participation will help in mobilizing human resources in respect of labour, managerial talents and political support which are crucial to programme success.
3. Participation will lead to have more sustainable programme even without outside funding or even without outside technical support.
4. Participation will lead to motivation of people for action and in creating a sense of belongingness among the people about the programme.
5. Participation of people provide not only new ideas for planning but also helps in accepting new ideas by people when put forth by trusted peers.

It is imperative that extension and development programmes are planned and executed not by the extension/ development personnel alone or the beneficiaries alone, but jointly by both. The people in the villages form the bulk of target groups, our extension efforts are targetted to. It is logical that a sincere desire and concerted efforts to have active participation of beneficiaries in all phases of development work are accepted as prerequisite for promoting rural development. Unless an extension and development worker accepts this premise, he will not be able to achieve the long term objectives of helping the people to help themselves.

The involvement of people in planning and execution of extension/development programme is important for the following reasons:

1. It develops a trusting atmosphere through discussions and deliberations with ultimate beneficiaries.
2. It provides a means to legitimize the programmes, their objectives and methods.
3. It helps in identifying local problems and their cause effects relationships, formulation of specific objectives and fixing priorities.
4. It ensures that the programmes are in tune with local current needs and interests of the people.
5. It helps determine and mobilise local resources, men, money and materials.
6. It facilitates identification and mobilization of actual and potential leaders.
7. It opens the way for gaining confidence, support and commitment of the people on a personal basis to carry out the programmes. The resulting programme becomes "ours."
8. It provides an opportunity to the participating members to develop leadership skills and competencies in the process of change.

9. It provides a good learning to the people in planning and execution of extension programmes through involvement in collecting and analysing facts, selecting problems for attention, assisting in carrying out agreed upon actions and in evaluating results.
10. It provides continuity in extension programmes when there is a change of extension personel. The people go ahead with the programe even in the absence of extension worker.
11. It enables the people acquire knowledge, develop skills and change attitudes for the development.
12. It develops self-confidence, a sense of responsibility and capability for critical thinking and problem-solving on the part of the beneficiaries.
13. It provides an opportunity to many people to satisfy a basic need to render service and to achieve status and recognition.
14. The public is better informed about the programmes and their outcome.
15. It brings the people in contact with sources of information and increases the ability to utilize the services of different organizations.
16. It results in better decisions as compared to those made by extension and development professionals alone.
17. It speeds up the process of educational change by multiplying contacts and efforts of extension personnel.
18. It ensures that no segment of the target clientale or geographic area that should be reached is excluded from the programme.
19. People feel motivated to cooperate in the

implementation of a programme in which they are involved in its planning. It provides a sense of commitment among beneficiaries.

20. In a democratic and free-choice society like ours, people have the right to decide for themselves which goal they will work towards.

The Process of Participatory Planning

The process of participatory planning is initiated after collecting, sharing and analysing all the informations by villagers using different PRA techniques. A group meeting preferably round table, is arranged for participatory planning on a convenient date and place where all concerned can willingly and leisurely participate. The administrators, the scientists and technocrates, the extension workers and beneficiaries should be invited and involved in participatory planning. However it is not possible to include all beneficiaries and even administrators and scientists for this work, nor it will be advisable to have a large group. It would be advisable to select the most concerned, knowledgable and well informed incumbents from each category. Such people should be invited in the meeting whose decisions are followed and listened by the people at various levels.

The place of meeting for the planning should be comfortable and free from all distraction and disturbances. It should be peaceful, calm and quite but with all required materials and resources. Adequate time should be kept for meeting. The process of participatory planning has the following steps :

1. Arrange to display all the outputs of different PRA exercises (maps, matrix, seasonal diagram, charts,

etc.) on the walls of the room where meeting is to be held.

2. Arrange the meeting at the predecided and pre-arranged common place preferably in the village setting.
3. Allow the villagers to look at their own analysis and drawings prepared by themsleves during the PRA exercise.
4. Arrange a presentation of all information by selected villagers to the large group.
5. Allow sufficient discusson on various aspects of village profile. Ensure participation of all villagers in discussion.
6. Ask the villagers to identify the problems of their village. While they name the problems, ask any one villager to write them on small pieces of paper. Keep the pieces of paper one by one vertically on the ground.
7. After an elaborate discussion on problems and collection of as many problems as possible, ask the villagers to arrange them in descending order of their priority.
8. Allow some time to prioritise the problems by villagers.
9. If one or two villagers are not in agreement with others, tell them that there would be further discussion on his particular point later.
10. Now, initiate discussion on the problems - where and when these problems have occurred.
11. Ask the villagers about how many villagers are affected with each of the problem and also what parcentage of loss it causes.
12. Develop matrix for computing priority status of

each problem on the basis of its importance, no. of villagers affected and percentage of loss.

13. Prioritize the problems on the basis of matrix scoring.
14. Now take the first priority problem and initiate discussion.
15. Ask the villagers to give the possible causes of the first problem. Write them in order in a chart form. Write socio-economic causes on the left side of the problem and bio-physical causes on the right side of the problem.
16. Ask the secondary causes (causes of a cause) and tertiary causes and arrange them in the chart/ diagram.
17. Prepare problem - cause diagram following the above procedure.
18. Take a problem cause diagram and discuss in detail with the villagers.
19. Ask the villagers to identify the causes which could be solved by them within their resources or with little outside help. Mark them.
20. Take one cause and ask the villagers about the possible solutions.
21. Remember all problem causes may not have development solutions/interventions.
22. Ask the villagers to rank the solutions/intervention in order of easiness and convenience. Decide the interventions following this procedure.
23. Move further ahead and have discussion on who could be the first beneficiaries to implement interventions or test interventions at their farm. Select the beneficiaries for each intervention to reduce a cause and to solve the problem.

24. List the inputs/help required from outside to implement the intervention with the selected beneficiaries. Quantify them.
25. Discuss the beneficiaries and make a programme for intervention.
26. Proceed to the next cause which could be reduced to solve the problem, and so on.
27. Proceed to the next problem and follow the same procedure for identifying causes and intervention and preparing intervention plans.
28. Assign the responsibilities on incumbents for implementation of the plan and seek commitment.
29. Prepare a time schedule for plan implementation in the form of on - farm- research, on-farm-trial, demonstration and application of intervention.
30. Also prepare a definite and time bound plan for resource mobilization and arranging required inputs for plan implementation.

The participatory planning workshop should be ended with definite product in the form of a realistic plan. All participants should feel a sense of satisfaction with definite commitment and goal in mind.

Identification and Prioritization of Problems

Farmer participation in agricultural technology development and refinement follows the sequence of Agro-eco system analysis, identification and prioritization of problems, participatory planning, problem cause diagrams, site characterization and identification of intervention points.

A participatary prioritization based on the criteria of regional productivity losses helps in ranking the problems. Prioritization is best carried out through active participation of farmers, extensionists, researchers etc. At this stage, problem that have technological solutions are transferred to the extension system for programming. The examples of prioritization of problems for Dohna, Manda, Kamua and Purenatal villages of Bareilly district have been presented in Fig. 21 to 24.

Problem identification is the joint responsibility of farmers, extension workers and scientists. Problems of villagers in respect of farm enterprises are best known and perceived by the respective farmers. Accordingly, the villagers' participation in problem identification is essential. Further, it is necessary that villagers from all sections related to the subject/enterprise must be included in seeking participation. For example, identification of problems in paddy cultivation in a village must include all types of paddy farmers, small, marginal and large, resource poor and resource rich, having problem and good and fertile lands with different source of irrigation.

Sometimes the villagers / farmers may not perceive the exact problem due to lack of knowledge and lack of proper information. Under these situations, the extension workers or scientists should feed the farmers with the relevent information so that they could perceive the problem. Once all problems are listed and prioritized through participatory planning, each problem starting from most priority need are required to be analysed critically one by one.

Problem-Cause Diagram

Problem-cause diagram represents all possible causes, primary, secondary and tertiary of a given problem organized in sequence and arranged in respect of bio-physical and socio-economic causes. The problem cause diagram are prepared through participatory planning. The farmers are asked to list

Fig. 21 : Prioritisation of Major Problems

Village: Dohna   No. of families: 250

Sl. No.	Problems	Areawise distribution of problem	Importance of Enterprise	Seriousness of problem	No. of families affected	Relative Importance of the problem
1.	Low Yield of Sugarcane	XXX	XXXX	XXXX	194	II
2.	Low yield of Paddy	XXXX	XXXX	XXXX	213	I
3.	Low yield of Wheat	XXX	XXXX	XX	189	IV
4.	Low yield of Pulses	XXX	XXX	XXX	60	IX
5.	Low milk yield of cattle and buffalo	XXX	XXX	XXXX	181	V
6.	High disease prevalance in live-stock	XXXX	XXX	XXXX	169	III
7.	High mortality of buffalo and cattle calf	XX	XXX	XXXX	140	VI
8.	Low conception rate of A.I.	XX	XXX	XXX	85	VIII
9.	Poor economic condition of women	XX	XX	XX	145	X
10.	fodder scarcity	XXX	X	XXX	181	VII

Legends: XXXX - Very important, XXX - Important, XX - less important, X - least important   Village facilitator: Group of farmers

Fig. 22 : Prioritisation of Major Problems

Village : Manda   No. of families : 135

Sl. No.	Problems	Area wise distribution of problem	Importance of problem	Seriosness of problem	No. of Affected families	Relative importance of the problem
1.	Low Yield of Sugarcane	XXXX	XXXX	XXXX	88	I
2.	Low yield of Paddy	XXX	XXX	XXXX	82	II
3.	Low yield of Wheat	XXX	XXXX	XX	78	VII
4.	Low Yield of Pulses	XX	XXXX	XXXX	53	V
5.	Low milk yield of cattle and buffalo	XXX	XXX	XXXX	79	IV
6.	High disease prevalence in live-stock	XXX	XXX	XXXX	83	III
7.	High mortality of cattle and buffalo calf	XXX	XX	XXX	73	VIII
8.	Low conception rate of A.I.	XX	XX	XXX	49	X
9.	Poor economic condition of women	XXX	XX	XX	52	IX
10.	fodder scarcity	XX	XX	X XXX	79	VI

Legends: XXXX - Very important, XXX - Important, XX - less important, X - least important

Village facilitators: Group of farmers

Fig. 23 : Prioritisation of Major Problems

Village: Kamua No. of families: 98

Sl. No.	Problem	Area wise distribution of problem	Importance of Enterprise	Seriousness of problem	No. of Affected families	Relative Importance of the problem
1.	Low Yield of Sugarcane	XXXX	XXXX	XXX	73	I
2.	Low Yield of Paddy	XXXX	XXXX	XXX	64	II
3.	Low yield of Wheat	XXX	XXX	XX	62	VI
4.	High pest and disease prevalence in Mango	XXXX	XXXX	XXXX	25	III
5.	Low Yield of Pulses	XX	XX	XXX	21	X
6.	Low milk yield of cattle and buffalo	XXX	XXX	XXX	61	V
7.	High disease prevalence in live-stock	XXXX	XX	XXXX	57	IV
8.	High mortality of cattle and buffalo calf	XXX	XX	XXXX	48	VII
9.	Low conception rate of A.I.	X	XX	XX	46	XI
10.	Poor economic condition of Women	XX	XX	XX	45	IX
11.	fodder scarcity	XX	XX	XXX	61	VIII

Legends: xxxx - Very important, xxx - Important, xx - less important, x - least important

Village facilitators: Group of farmers

Fig. 24 : Prioritisation of Major Problems

Village : Purenatal No. of families : 182

Sl. No.	Problem	Area wise distribution of problem	Importance of Enterprise	Seriousness of problem	No. of affected families	Relative importance of the problem
1.	Low yield of Sugarcane	XXXX	XXX	XXXX	125	II
2.	Low yield of Paddy	XXX	XXX	XXX	125	IV
3	Low Yield of Wheat	XXX	XXX	XXX	116	V
4.	High disease and pest prevalence in Mango	X	XX	XXX	12	XII
5.	Low yield of pulses	XX	XX	XXXX	48	VIII
6.	Low milk yield in cattle and buffalo	XXX	XXX	XXX	80	VI
7.	High mortality of cattle and buffalo calf	XX	XX	XXX	101	I
8.	Low conception rate of A.I.	XXXX	XXX	XXX	117	VII
9.	High disease prevalence in live-stock	XXXX	XXXX	XXXX	126	III
10.	Poor economic condition of women	XX	X	XX	64	XI
11.	fodder scarcity	XXX	XX	XX	80	IX
12.	Infertility in cows	XX	XXX	XXX	45	X

Legends : XXXX-Very important, XXX-Important, XX-Less important, X-least important

Village facilitator : Group of farmers

all possible causes of the problem.

All the possible collected and listed causes are classified into two categories, viz. socio-economic and bio-physical. The socio-economic causes are arranged on the left side of the problem and bio-physical causes arranged on the right side of the problem. The causes are also classified into primary, secondary and tertiary causes along with their interrelationships. The primary causes are arranged near the problem. The secondary causes are arranged in second line and in line with respective primary causes. Tertiary causes are arranged in the third line accordingly. The relationship between problem and causes and between primary, secondary and tertiary causes are depicted by connecting arrows. The relative importance of each cause may also be depicted by the width and size of the block where in the cause is written.

Example : 1

Problem - Low Yield of Sugarcane

Fig. 25 presents a problem cause diagram drawn by the farmers of Purenatal village on their one of the priority problem - low yield of sugarcane. Socio-economic causes identified by the farmers have been placed on the left side of the problem block while bio-physical causes have been placed on the right side. Primary socio economic causes included lack of knowledge and skills and antilope problem. The primary bio-physical problems included imbalanced use of fertilizer, poor plant population, poor weed management, high incidence of pests and diseases, lodging of sugarcane and poor ratoon cultivation. The secondary causes affecting the primary cause imbalanced use of fertilizer are non-availability of fertilizer at proper time, lack of knowledge and skills, adulterated fertilizer and lack of credit facilities. Similarly, the secondary causes for the primary cause poor plant population are poor quality seed, late planting and lack of

Fig. 25 : Problem-Cause Diagram of Low Yield of Sugarcane Crop

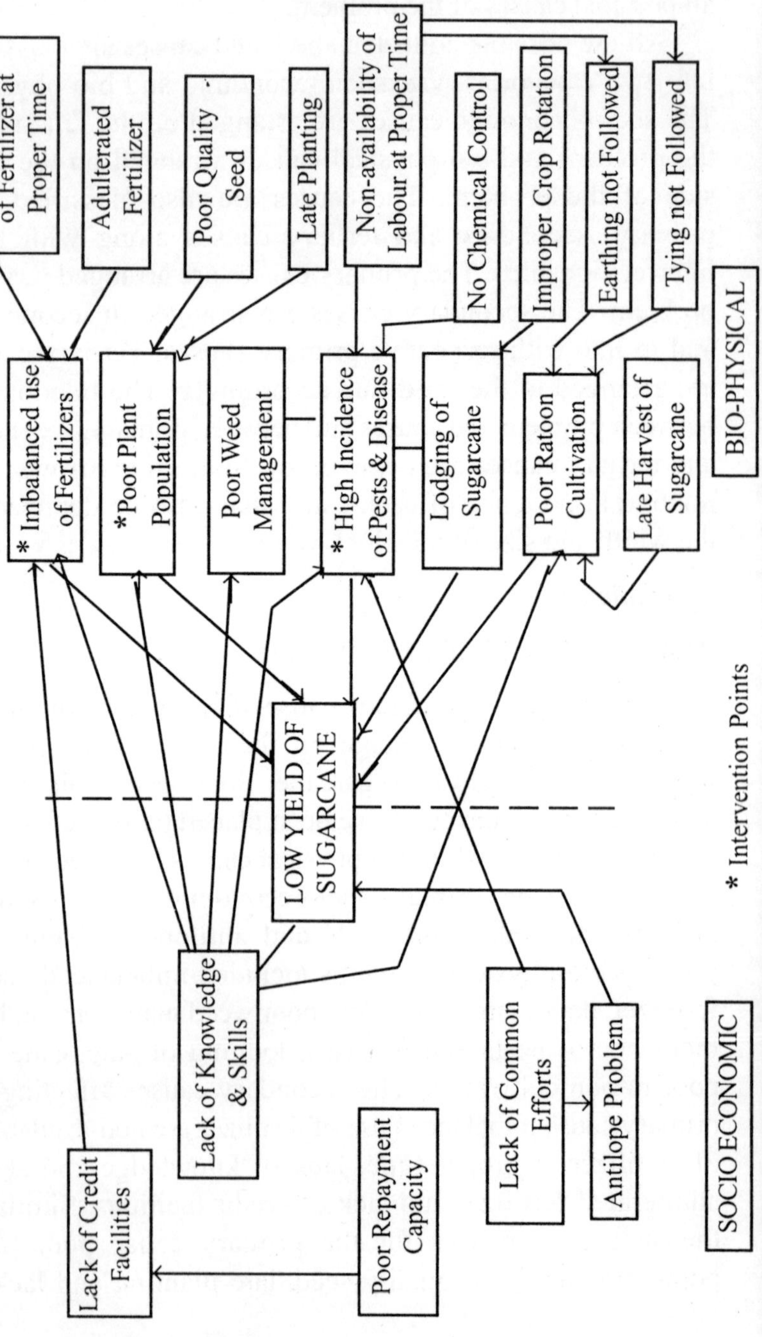

knowledge and skill. All the secondary and tertiary causes are placed at proper places. The diagram is made step by step in order to give a clear and presentable shape. The secondary and tertiary causes provide clues for practical and feasible interventions. After drawing the problem cause diagram the farmers are asked to select a few most important causes on which work could be initiated conveniently within the available resorces or with little external input. The causes which are difficult to resolve are left as such. Farmers of Purenatal village identified three such intervention points, viz. imbalanced use of fertilizer, poor plant population and high incidence of pests and diseases. The intervention points were identified keeping in view of all the site characterization of the problem which gives critical analysis of the problem. Finally, intervention programme was prepared on these selected intervention points.

Example 2 :

Problem-Low Conception Rate of Artificial Insemination

Fig. 26 presents a problem cause diagram drawn by the livestock owners of Purenatal village on their one of the priority problem low conception rate of artificial insemination.

Keeping the problem block in the middle the socio-economic causes are placed on the left side which included lack of knowledge and skills, poor economic condition and high cost input. But, none of these are primary causes. The primary causes (bio physical) are - poor heat detection, poor quality of semen, frequent reproductive disorder, lack of specific breed semen, high disease occurance in crossbred, imbalanced feeding and poor management. The high disease occurance was reported due to several secondary and tertiary causes which has been shown by arrows. These are poor veterianary services which is due to far distance of veterinary hospital and more operational area of the hospital. The other

Fig. 26 : Problem-Cause Diagram of Low Conception Rate of AI

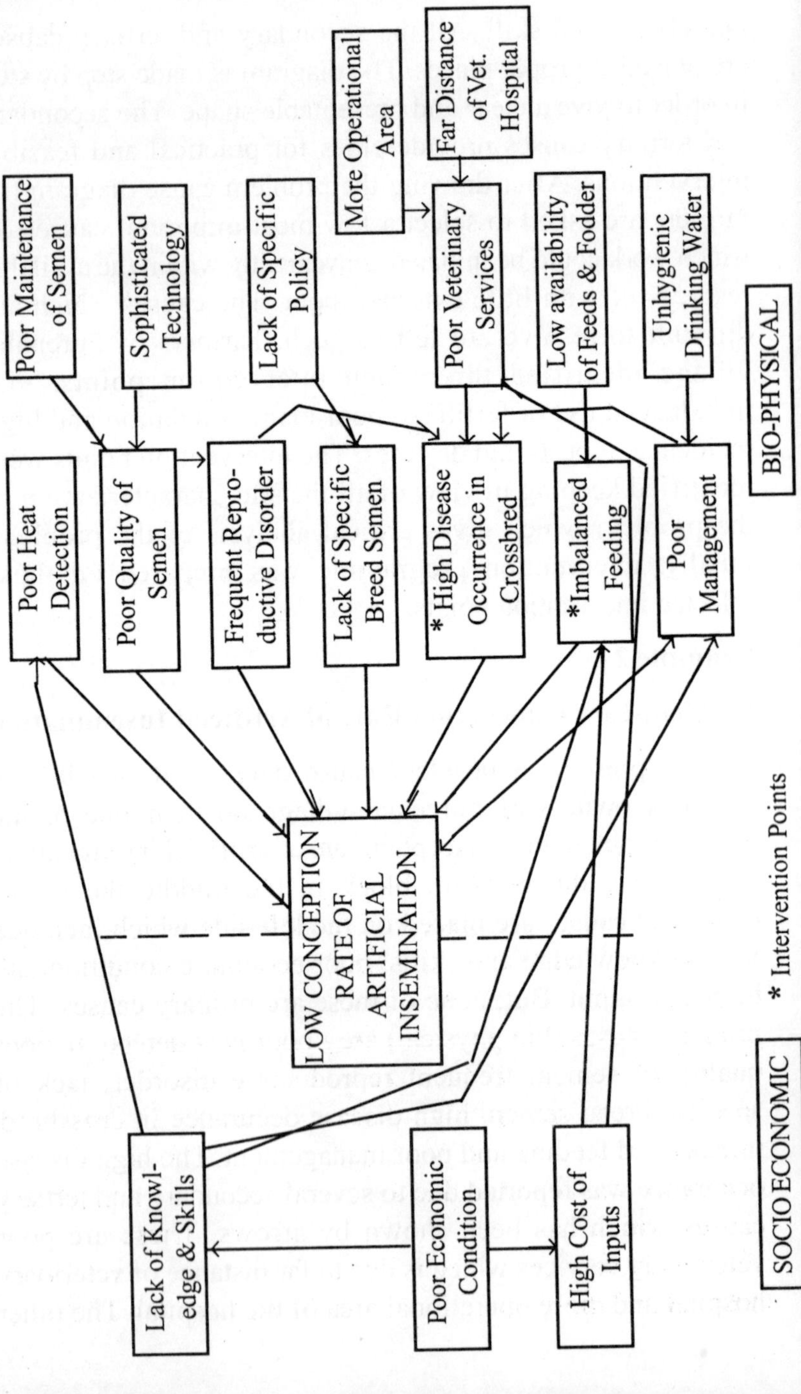

secondary causes like imbalanced feeding and poor management are due to lack of knowledge and skills, low availability of feeds and fodder, high cost of inputs and poor economic condition of the farmers.

The important intervention points selected by the livestock owners on which programmes were developed are: high disease occurance in crossbred, imbalanced feeding and lack of knowledge and skills. These intervention points were identified keeping in view the site characterization of the problem which presents the critical view of the situations under which the problem exists.

Site Characterization

Site Characterization is an important step in developing the intervention programme for solving the existing problems. The concerned members of the participatory planning group list out the site characteristics in respect of specific problem. This include how the problem could have originated, what could be the situations and factors responsible for the problem? Who are the victims of the problem and who are not? What could be the possible barriers in solving the problem? What is the intensity of the problem and how much loss it causes? The site characterization also include the socio-economic and psycho-cultural variables/situation which affect the problem and its solution at the village and farmers level.

Site characterization helps in identifying solutions from among the existing resources. This avoids outside search and even conducting new research and hence saves lots of time and expenditure. Only those problems which do not find solutions from among the available alternatives and resources need to be looked for searching solution either from outside or through conducting research on the specified problems. The solutions available outside system need to be examined critically in accordance with the site characterization for their extrapolation. Such matching solutions need to be tested in the system.

Site characterization for a problem related to crop production must detail out the following aspects :

- Soil type
- Land use pattern
- Climate - seasonal variations in temperature, rainfall and humidity.
- Irrigation and source of irrigation
- Cropping pattern
- Production and productivity status of respective crop.
- Major insect pests in the respective crop.
- Major varieties of the crop grown
- Cultural practices and crop calender
- Input availability - money, manpower, material, seed, fertilizer, pesticide, etc.
- Recommended practices, their adaptability and adoption status.
- Cost of production
- Marketability of produce and price status
- Profitability status
- Major problems in order of priority
- Historical perspective and causal analysis of the problem.

Site characterization helps in identifying intervention points on which priority action could be initiated. It outlines the situations where the possible interventions and related alternative solutions could be tested. It provides a platefarm for extrapolation of test results. The site characterization should be in as much detail, as possible but specific in respect of the problem. The description should be in logical sequence and directional towards the problem and its possible solutions and interventions.

Example 1:

Site characterization in respect of the problem of Low Yield of Sugarcane in Purenatal village of Bareilly district

Sugarcane is the first major crop in Purenatal village. A total area of about 1000 bigha is covered under sugarcane including ratoon crop. Almost the entire area of sugarcane is under improved varieties with high production protentiality. But, the yields are low.

As many as 125 farmers from Purenatal have indicated the problem of low yield of sugarcane. The major causes of this problem identified by the villagers are imbalanced and inadequate use of fertilizer/manure, high incidence of pests and diseases, poor plant population, poor weed management, lodging of standing crop, poor quality seed, late planting, improper crop rotation, etc.

The farmers belonging to small production system category face the major problem of money availability for purchase of adequate dose of fertilizer and pesticides. Many farmers having small piece of land do not solely depend on farming and have to have other livelihood enterprise such as tonga riding, earth work outside the village, labourer on other's fields, etc. Such farmers plant the sugarcane crop on their small piece of land and leave it on the mercy of God. They practically do not have time for its care as they are heavily engaged in earning their livelihood through other enterprises. Such small and marginal farmers depend on the large and medium farmers with whom they work as labourer even for sowing of their own crop.They are able to sow their sugarcane crop only when they are spared for it by their employer - large farmers. Moreover, they have to use the plough and bullocks /tractor of those large farmers. Hence, even the sowing of sugarcane is delayed in their fields. After sowing of the crop, such farmers are not able to take care for hoeing, weeding, earthing, manuring, top dressing, irrigation, spray and dusting for protection from insects and pests, etc. They leave the crop as such and come to the field only for harvest. Whatever they get, they are satisfied. After all they do not depend on sugarcane for their livelihood.

On the other hand, the medium and large farmers, who depend much on sugarcane production enterprise face the problem of timely availability of inputs. Adequate manure is not available in the village. They face the labour problem also at peak seasons for which the operations are delayed. More specially the sowing is delayed. Many farmers sow sugarcane after the harvest of mustard or lentil crop. They lack in knowledge in respect of control of insects and pests and balanced use of fertilizer, etc.

Example 2 :

Site characterization in respect of the problem of Low Conception Rate of Artificial Insemination in Purenatal village of Bareilly district

Artificial Insemination technique was perfected as early as 1940 s and as a policy only A.I. is to be followed both in cattle and buffalo for breeding to solve the problem of inadequate availability of pedigree bulls. All the stray bulls are curled and rural livestock are provided with A.I. services. But, in most of the villages, especially in remote areas, the A.I. has not been so successful. The conception rate is as low as 10% at many places. Moreover, the State Animal Husbandry Department is not able to provide effective A.I. services to all the villages due to number of limitations. Further, inadequate and imbalanced nutritional status, reproductive disorders and lack of farmers knowledge also contribute to the failures of A.I. services. This is a very serious problem in view of the fact that as a policy, natural service has been banned and maintenance of bulls including stray bulls left earlier by the farmers have been discouraged and there is nothing to inseminate the cows/buffloes coming in heat.

The problem was much more felt serious in Purenatal village where the farmers have livestock specially dairy production as an important enterprise. This problem was ranked as III by the farmers of this village. Further, it has also been reported that the problem is more serious in case of cross-bred cattle and that the farmers being progressive and innovative in this village adopted cross-breds and keep in good number. In all 17 livestock owners are in loss in the village due to this problem. Many of the farmers maintain 5 to 20 crossbreds in the village. Veterinary and artificial insemination services are not readily available in the village. The animals are not able to receive timely services. Insemination done untimely can not yield results. In case of reproductive disorders, timely diagnosis and treatment by qualified and expert veterinary doctor is not available. Continuous lack of proper treatment led to infertility in cattle.

Another reason for low conception rate could be imbalanced feeding. The cross-breds are maintained by resource rich farmers who have temptation of higher milk yield and feed the animals even more fodder and concentrate than requirement. But, due to lack of knowledge do not provide balanced feeding. Fodder lacks in green fodder specially during summer season. In the name of concentrate mostly the food grain like wheat is given exclusively. This lacks in oilseed cake and Dal chuni which are rich in protein. Mineral mixture is very rarely given.

Animals are kept in poorly managed sheds except in few

cases. The sheds are mostly katcha and uncleaned. They do not have proper drainage for flow of urine and cow dung. The sheds and animals are not disinfected regularly. Animals are not dewormed and might be having high parasitic load.

The major causes of low conception rate of A.I. are imbalanced feeding, high disease occurance and frequent reproductive disorders, poor heat detection, poor A.I. services and veterinary services, sophisticated technology, lack of cold chain for storing and transporting semen with State Government, lack of poor management practices by the farmers, lack of knowledge and skills and poor economic condition of the livestock owners.

Identification of Intervention Points

The problem - cause diagram prepared through participatory approach indicate several causes of a given problem. Each cause is analysed in respect of possible solution, its intensity and easiness in solving it. The cause which could be tackeled easily could be taken first. These are called as intervention points. Participatory planning meeting discusses and analyses each cause critically. All possible solutions for a specific cause are listed. From the alternative solutions and technologies, the most feasible one is selected keeping the farmers situation in mind for implementation.

Technology Assessment and Refinement

The selected solution and technology are analysed in respect of actions to be taken in the form of modules. The technology modules could be subjected to on-farm-research, on-farm-trial, demonstration, training, farmers' visit to research centre, consultancy by experts, etc.

Decision needs to be taken through participatory planning approach about the above mentioned modules, viz. OFR,

OFT, demonstration, training, etc. in respect of the farmers to be involved in each for a given technology or intervention point for its assessment and refinement. An exampple of decision analysis is given below through PRA method for the problem of 'low yield of sugarcane' in the four selected villages:

Example :

Problem : Low yield of sugarcane
Problem cause diagram: Fig. 21
Intervention points :

1. Imbalanced use of fertilizer
2. Poor plant population
3. High incidence of pests and disease
4. In proper selection of vareity.

Technology Interventions

For Small Production System:

1.	Balanced use of fertilizer in sugarcane under limited resource conditions	OFR
2.	Proper selection of varieties of sugarcane under limited resource onditions	OFR
3.	Maintenance of optimum plant population and control of termite and stem-rot disease	Demonstrations
4.	Control of top shoot borer in sugarcane	Demostrations

For Green Revolution Production System:

1.	Balanced use of fertilizer in sugarcane under better resource conditions	OFT

Technology Modules

TM-1: Balanced use of fertilizer in sugarcane under limited resource conditions.

On Farm Research

Treatments : T-0 Farmers' practice (50:15:0)
T-1 Bio-fertilizer+NPK (50:15:15)
T-2 Bio-fertilizer+NPK (75:30:30)
T-3 Bio-fertilizer+NPK (100:45:45)

Plot Size : 800 sq.m. (200x4)

Replications : 40 (in four villages)

Critical inputs : (i) Bio fertilizer
(ii) Fertilizer - Urea
- DAP
- MOP

Cost of each OFR : 150.00

Total Cost : 150 × 40 = Rs. 6000.00

TM-2: Maintenance of optimum plant population and control of termite and stem rot disease in sugarcane under limited resource conditions.

Demonstration

Treatments : T-0 Farmers' Practice (r × r = 75 m + no chemical control)
T-1 Use of Folidol dust + seed treatment + proper plantation method (r × r = 90 cm)

Plot size : 1000 sq. m. (500x2)

No. of Farmers : 40 (in four villages)

Critical inputs : (i) Folidol dust for termite control.
(ii) Aration for seed treatment to control red rot disease.

Cost of each demonstration : 50.00

Total Cost : 50 × 40 = Rs. 2000.00

TM-3 : Balanced use of fertilizer in sugarcane under better resource conditions.

On-Farm-Trial

Treatments:	T-0 Farmers' practice (75:30:0)
	T-1 Bio-fertilizer+NPK (100:40:40)
	T-2 NPK (150:60:60)
Plot Size :	900 sq.m. (300x3)
Replications :	40 (in four villages)
Critical inputs :	(i) Bio-fertilizer
	(ii) Urea
	(iii)DAP
	(iv) MOP
Cost of each OFT	: 310.00
Total Cost	: 310 × 40 = Rs. 12,400.00

Farmers and farmers' plots under location specific conditions are identified in the village through participatory planning process for implementing OFR, OFT, Demonstration, etc. Farmers' commitment for implementing the technoogy assessment and refnement process is obtained alongwith agreement to contribute in respect of his share of involvement like land, water, labour, etc. Critical inputs are arranged well in time and experiments are conducted under the expert guidance and supervision. Adequate care needs to be taken to maintain the honesty and sanctity of the experiments. Enough safety arrangements may be made against any damage to the experiment. Critical operations/ activities of the experiment particularly the treatments should be done in the presence of all concerned and also in presence of as many farmers as possible. The results/observations should be taken and recorded timely. The results could be analysed critically and statistically wherever necessary. Farmers' perceptions on each observation may also be collected and recorded.

The results of experiments (OFR and OFT) will provide the assessment of technology in respect of either conforming it or refining it to suit the local conditions. The results of demonstrations will provide further confirmation and develop confidence. The results will be able to convince the farmers for adoption.

On-Farm-Research (OFR)

On-farm-research (OFR) is the process of experimentation to test and evaluate the economic and bio-physical viability and social acceptance of an improved technology or practice. OFR is intended to reveal effects of experimental treatments under location specific situations. The emphasis is on evaluation of new technology from socio-economic perspectives. On-farm- research are designed not only to assess and verify the technology but also to refine and modify it to suit the local situations. For instnace the recommendations of fertilizer dose in a given crop is assessed and tested for modifications on specific soil in specific variety for a specific farm and farmer, since same dose will not be applicable under all situations. Similarly, feeds and fodder need to be experimented on local animals for their effect before recommending the specific ration to a given livestock.

In real sense, OFR is a full research on the farmers' field which provides specific recommendation for the given situation. These recommendations generally will not be applicable for different situations. OFR normally requires analysis of statistically significance of differences. These are conducted where it could be assumed that the recommendations made by on-station-research will not be as such applicable and need to be modified, may be in respect of dose, may be in respect of varieties or may be in respect of insecticide or pesticide. The recommendations could be different for small and marginal farmers. These would vary in accordance with soil type, fertility of soil, level of irrigation availability, etc.

On-Farm-Trial (OFT)

On-farm-trial could also be called as verification trials where the on-station technology is as such verified for its applicability and efficiency on the farmers fields of almost similar situations. OFTs may not be having dose variations but could still have more than one treatment. The treatments could be the input variations. OFTs are done when it is almost certain that the technology will make an effect and will be viable. Examples of OFT could be the test of efficacy of vaccination against any epidemic in livestock. Under these situations, it is almost sure that the vaccination will produce immunity in the animal against the specific disease. Under such OFTs, the dose can not be changed. The technology can not be modified under different socio-economic situations. There could be input variations and therefore different inputs/technologies could be tested from their cost-benefit point of view, so that specific recomendation could be made under specific socio-economic conditions.

Demonstration

Demonstration is the process to show the worth of a practice or technology. Demonstraton is conducted where we are sure that the given technology would certainly be effective and feasible under given situations. Normally, demonstrations are conducted for such technology which has already been tested for its efficacy and viability under the given situations. Demonstration is therefore the next step to OFTs or OFRs. The major purpose of demonstraton is dissemination of technology. The objective of demonstration could, however, also be to confirm the result of the technology. This develops confidence among the technology promotors as well as technology users. Demonstration follows the principle of 'seeing is believing'. This has been considered as the best extension teaching method since it has the highest power of convincing the farmers and leads to adoption.

Demonstration could be conducted for any type of technology be it tested through on-farm-research or through on-farm-trial. It is conducted to show the worth and superiority of the technology over the existing practices. It is therefore, necessary to maintain records in respect of all attributes where superiority is claimed while conducting demonstration. Demonstration may also consist of more than one treatments, not only in respect of use of different inputs but also in respect of different doses depending upon the type and nature of technology. Demonstration also need to be conducted under different location specific situations and under different socio-economic condition.

9
Technology Dissemination and Technology Integration

Technology Dissemination

The technologies tested and refined through location specific OFR and OFTs may be considered ready for dissemination among farmers for adoption. The dissemination process of technology could start right from the conduct of OFR, OFT and Demonstrations. The experimenting and demonstrating farmers will play an important role in disseminating the technologies. The results would provide specific direction in respect of suitable clients for communicating the technology where it could yield better results.

Localised and personalised communication channels could be more effective in disseminating the tested and refined technology in the specified area. Mass media like radio and television may not be so appropriate. Meetings, discussions, demonstrations, field visits, case presentations, local exhibitions, etc. could be used effectively. The participating farmers in OFR and OFT could be involved in disseminating the technology.

The large number of replications for each OFR, OFT and Demonstrations will make the disseminations more easy. The use of technology could be made easy and it is possible that the entire village is covered with the recommended technology within the next one season. The technology assessment and refirement will itself be an effort for disseminating the technology. In case, however, the technology is not proved worthwhile through OFR and OFT, it will be rejected at the point itself for ever. In view of this very little efforts for disseminating the technology will be required in the villages where OFR/OFTs have been conducted.

Training

The technologies which are known to be superior and are confirmed for their viability under the prevailing situations are required to be applied in the field at the farmers' plots. The new technologies, however, differ from the existing ones in respect of inputs, operations and process for which the farmers need training. The farmers might be convinced about the technology by observing the results at research stations or through OFR/OFT/Demonstrations, but do not have sufficient knowledge and skill for making use of the technology. Under such situation, training needs to be arranged for farmers as an important intervention. Training needs to be given either for providing knowledge or for developing skills or for both.

The decision about the training could be taken during the participatory planning meeting. The farmers can identify their own specific training needs and the number of farmers to be trained. The scientists and extension professionals may help in identifying the training organization where the farmers can get training. Specific plans may be prepared for training.

Both on-campus as well as off-campus training programmes could be arranged for farmers based on the requirement of the training programme as well as on the conveniences of the farmers. The training programmes could be arranged in batches based on the requirements as how many farmers need training.

Farmer's visit to Research Centre

Farmers' visits are arranged to research centres both for convincing the farmers through confirmation of supperiority of the technology as well as to make them known about the process involved in the use of technology. Farmers' visit to research centre provides reinforcement to farmer's learning. It provides further opportunity to the farmers to clarify if they have any doubt on the efficiency and use of technology. Normally the resource-rich farmers and those having the similar situations as that of the research stations are taken for the visit to research stations for obtaining better results. The visit of resource poor farmers and those having different situations will not have desirable results.

Consultancy by Experts

Farmers, who would be using the technology for the first time on their farm need to be provided constant guidance and consultancy by experts. The experts need to pay visits to the farmer's fields specially at the time of using the technology. Any mistake committed by farmers in using the technology will lead to failure of technology, which will have adverse effect on the farmers' confidence and will lead to frustration and non-adoption. Such non-adoption of one technology will further lead to non-adoption of other technology and loss of faith towards extension workers and scientists.

In view of this, this last step in technology dissemination and adoption is very critical and important. The experts' visit to the field at critical occasions, therefore, must be arranged in such a way that the farmers get corrective guidance in using the technology.

Tecnology Integration

Assimilation of technology in the normal process of production activities may be called technology integration. Technology once adopted by the farmer will have to be assimilated with the normal activities at the farm. Technology in this process looses its identity and is converted into practice. Adoption of technology is converted into habit of the farmer, farm and farm family. Technology integration makes the balances and provides a set place for the technology among all the elements and process of productions.

There are number of examples when the tehnology though seem to be sound but could not be sustained under a given system and thus could not be integrated. Cross breeding in cattle could not be integrated in the system specially in UP and particularly around Bareilly. Though the technology is sound enough to increase milk production failed to be integrated due to several reasons such as cross breds require regular and constant care by veterinary doctor which is not available in the villages, its milk is not liked by the people and thus failed to get suitable market and price and its bulloks are not so active for agriculture as Desi ones. Hence, the technology could not be integrated in the system even in 50 years.

The technology needs to be integrated in the total production system. Unless it is integrated in the system, it remains isolated and can be thrown out by the system at any moment of time. Such isolated technology remain and continue to be adopted till the outside support or force is maintained. Once, the support is withdrawn, the technology is discontinued. In view of this the technology needs to be integrated in the system.

Technology integration may require certain changes in other related elements of the system. Unless these changes are made, for which simultaneous efforts need to be made, the technology remains isolated which could be rejected at any time.

The process of technology identification, technology testing and technology dissemination must be looked into from the systems point of view under the specific agro-eco system. In fact, the basis of participatory approach lies on this principle. The entire process of technology development to the technology adoption must be viewed under the farmers' perspective and also under the systems perspective.

Bibliography

Anonymous. Pilot Project on Technology Assessment and Refinement through Institution - Village Linkage Programme (IVLP) Progress Report (1996-1997). Krishi Vigyan Kendra, IVRI, Izatnagar, 1997.

Anonymous. Pilot Project on Technology Assessment and Refinement through Institution-Village Linkage Programme (IVLP) - Progress Report. Presented in First National Workshop of the Project at IVRI, Izatnagar held on Feb. 3-5, 1997.

Anonymous. Farmers Participation in Agricultural Research and Extension Systems. NIRD, Hyderabad, Concept Publishing Co. 1997.

Anonymous. National Agricultural Technology Project (NATP) Concept Paper. ICAR and Deptt. of Agri. & Cooperation, Govt. of India, Krishi Bhavan, New Delhi, February 1995.

Anonymous. Pilot Project on Technology Assessment and Refinement through Institution Village Linkage. Project Document, Division of Agricultural Extension, ICAR, New Delhi, 1995.

Anonymous. Pilot Project on Technology Assessment and Refinement through Institution - Village Linkage Programme (IVLP) - Action Plan. Krishi Vigyan Kendra, IVRI, Izatnagar, 1997.

Anonymous. Village Entry, Community Forestry Working Paper-2, Prepared by Institute for Sustainable Development. Haryana Forest Department, Chandigarh, 2000.

Arulraj, S. Farmers as Experimenters in Participatory Technology Development. Lecture in winter school, Kerala Agricultural University, pp. 21-28, 2002.

Arya, H.P.S. and Jamal, Shagufta. Pilot Project on Technology Assessment and Refinement through Institution Village Linkage - Project Proposal, Krishi Vigyan Kendra, IVRI, Izatnagar, p. 78 + xxxvii, 1995.

Ashby, J.A. and Sperling, L. Institutionalising participatory, client - driven research and technology development in agriculture. Agricultural Administration Network Paper 49, p. 21, 1994.

Ashby, J.A. Institutionalising farmers, Participation in Agricultural Research. Annual Report of the Farmer Participation Special Project.

Colombia: CIAT, 1993.

Ashby, J.A. Methodology for the Participation of Small Farmers in the Design of On-Farm Trials. Agricultural Administration 22 (19) : 1-19, 1986.

Ashby, J.A. Participation of Small Farmers in technology assessment. IFDC-CIAT Phosphorus Project, Colombia, 1984.

Axinn, George, H. Guide on Alternative Extension Approaches. Agricultural Education and Extension Service. Human Resources, Institution and Agrarian Reforms Division, FAO. United Nations, 1988.

Benor, D and Baxter, M. Training and Visit Extension. The World Bank, Washington, D.C., 1974.

Biggs, S.D. Monitoring and Control in agricultural research systems on maize in Northern India. Research Policy 12 (1): 37-59, 1983.

Chambers, R. and Ghildyal, B.P. Agricultural Research for Resource Poor Farmers : The Farmer-First-and-Last Model. Agricultural Administration, 20:1-30, 1985.

Chambers, R. Participatory Rural Appraisal (PRA) : Analysis of experiences. World Development, 22 (9) : 1253-1268, 1994.

Chambers, R. Participatory Rural Appraisal (PRA) : Challenges, Potential and Paradigm. Institute of Development Studies, Sussex, 1993.

Chambers, R. Rapid Rural Appraisal : Rationale and repertone. Public Administration and Development, 1(2), 1981.

Chambers, R. Rural Development - Putting the last first. London, Longman, 1983.

Collinson, M.A low cost approach to understanding farmers. Agricultural Administration. 8(6) : 433-50, 1981.

Conway, G.R. Agro eco system Analysis. Agricultural Administration 20(1): 31, 1985.

Conway, G.R. Editorial-What's in a name. RRA Notes No. 1, 1988.

Conway, G.R.; Main, A.; Alam, Z.; Yar, M. and Husain, T. Agro ecosystem Zoning of the Hunza Valley: First Iteration. Draft Report. International Institute for Environment and Development, London, 1987.

Farrington, J. and Martin, A.M. Farmer Participatory Research : A Review of Concept and Recent Field Work. Agricultural Administration and Extension 29: 247-264, 1988.

Farrington, J. Organizational roles in farmer participatory research and extension: Lessons from the last decade. Natural Resource Perspective, 27, 1998.

Hilderbrand, P.E. Combining disciplines in rapid appraisal : The Sondeo Approach. Agricultural Administrations, 8(6) : 423, 1981.

Johnkuthy, I. Participatory Technology Development - Experience with the On-Farm Research Project, Lecture notes of the winter school, Kerala Agri. University pp. 92-99, 2002.

Kumar, B. and Hansra, B.S. Extenston Education for Human Resource Development. Concept Publishing Co., New Delhi, p. 200, 2000.

Maxwell, S. Farming Systems Research. Instt. Of Dev. Studies. Univ. of Sussex, Brighton, 1983.

Mettrick, H. Development Oriented Research in Agriculture, and ICRA Text book. International centre for Development Oriented research in agriculture. Wageningen, The Nether lands, 1993.

Mukherjee, N., Kumar, V.l Jayaswal, M. and Jena, B. Learning to Share-Experiences and Reflections or PRA and community Participation, Concept Publishing Co., New Delhi, 1998.

Mukherjee, Neela. Participatory Appraisal of Natural Resources. Concept Publishing Co. New Delhi, 1997.

Mukherjee, Neela. Participatory Rural Appraisal : Methodology and Applications, Concept Publishing Co., New Delhi, p. 160, 1993.

Mukherjee, Neela. Participatory Rural Appraisal and Questionnaire Survey. Concept Publishing Co., New Delhi, p. 163, 1945.

Mutsaers, J.W. and Walker, P. (eds.). On-farm Research in Theory and Practice. Proceedings of an IITA workshop on design and analysis of on-farm trials 27 Feb. - 3 Mar. 1989. Intec Printers Ltd. Ibadan, 1989.

Muylwijk, J.G. Gender Issues in Indian Livestock. Indo-Dutch Project on Bioconversion of Crop Residues. Wageningen, 1994.

Norman, D.W. The Farming Systems Research : Relevancy for the Small Farmer. Michigam State Uni. Rural Development Paper No. 5, Michigan, 1980.

Peter, K.V. Participatory technology-concepts and methodology. Key note address at the winter school on participatory technology development concept and methodology' organized on September 5-26, 2002at Collect of Agriculture, Padannakkad, Kerala Agricultural University.

Prasad, C. and Singh, S.P. Research Education-Extension-Clientele Linkages for accelerating farm productivity and agri-business. Paper summary Published in Extended summaries Vol. 1 of second International Agronomy Congress on Balancing Food and Environmental Security - A Continuing Challenge. New Delhi,

Ind. Soc. Of Agronomy. November 26-30, pp. 685-688, 2002.

Prasad, R.M. Concept of Participation - A Theoretical perspective. Lecture Notes of the winter school, Kerala Agril. University, pp. 12-16, 2002.

Prasad, R.M. PTD Framework and salient Features. Lecture Notes of the Winter School, Kerala Agril. University, pp. 17-20.

Pretty, J. Portraits and stories, RRA Notes No. 3, ILED, London, 1988.

Rhoades, R.E. Three models of the development and generation of acceptable agricultural technology. International Potato Centre, Peru, 1984.

Rhoades, R.E. and Booth, R.H. Farmer-back to farmer: A model for generating acceptable technology. Agricultural Administration, 11(2): 127-137, 1982.

Roling, N. Extension Science. Cambridge University Press, New York, 1988.

Shaner, W.W., Philipp, P.E. and Schmehl, W.R. Farming Systems Research and Development : Guidelines for developing countries. Westview Press, Colorado, 1982.

Singh, K. and Schiere, J.B. On-Farm Research for Testing of Appropriate Technologies in Crop-Livestock Production Systems. ICAR/BAIF/ BIOCON/WINROCK International. Technical Bulletin No. 3, ICAR, New Delhi, 1994.

Singh, S.P. Pilot Project on Technology Assessment and Refinement operational Guidelines (Mimeo). Div. of Agri. Extn., ICAR, New Delhi, 1995.

Sutherland, A., Martin, A. and Salmon, J. Recent experiences with participatory technology development in Africa : Practitioners Review. Natural Resource Perspective, 25, 1997.

Swaminathan, M.S. Science and integrated rural development, Concept Publishing Co., New Delhi, 1982.

Swaminathan, M.S. Science and the conquest of hunger. Concept Publishing Co., New Delhi, 1983.

Villiers, Anne K. de Quantifying Indigenous Knowledge : A Rapid method for Assessing Crop performance without Field Trials. Agricultural Research and Extension Network, Paper 66. Overseas Development Institute, London, 1996.

Waters Baifer, A. Participatory technology development in ecologically oriented agriculture : some approaches and tools. Agricultural Research and Extension Network, Paper 7: ODI: London, 1989.

Index